EASTERN TIMES

CONTENTS

Introduction	3
A Day to Remember	4-16
L.N.E.R. Locomotive Mileages and Availability in the War Years	17-23
Colwick – A forgotten great northern epicentre	24-37
The 'Alpine' Route – linking Halifax, Bradford and Keighley (part 1)	38-50
The West Riding	51
A funny looking engine	52-53
My Trainspotting Odyssey – 1958	54-60
No. 2395 Britain's mightiest locomotive	61-66
Eastern Region Totem Signs – the scarcest of them all	67-69
Deltic Days at Kings Cross	70-77
The Headshunt	78-80

The Transport Treasury

Standard 2MT 2-6-0 No. 78046 crosses the Royal Border Bridge at Berwick-upon-Tweed on 25th September 1961 with the 6.35p.m. to St. Boswells.
Photo: W. A. C. Smith © Transport Treasury

© Images and design: The Transport Treasury 2024. Design and Text: Peter Sikes

ISBN: 978-1-913251-76-5

First published in 2024 by Transport Treasury Publishing Ltd., 16 Highworth Close, High Wycombe HP13 7PJ.

The copyright holders hereby give notice that all rights to this work are reserved.
Aside from brief passages for the purpose of review, no part of this work may be reproduced, copied by electronic or other means, or otherwise stored in any information storage and retrieval system without written permission from the Publisher.
This includes the illustrations herein which shall remain the copyright of the copyright holder.

Copies of many of the images in EASTERN TIMES are available for purchase/download. In addition the Transport Treasury Archive contains tens of thousands of other UK, Irish and some European railway photographs.

www.ttpublishing.co.uk or for editorial issues and contributions email: tteasterntimes@gmail.com

Printed in the UK by Short Run Press, Bittern Road, Sowton Industrial Estate, Exeter EX2 7LW.

INTRODUCTION

Another edition and another collection of interesting and varied articles await the reader, all accompanied by captivating images. Issue 4 has the following in store for you.

Cleethorpes in the late 1950s and early 1960s is the focus of our first article. Railway enthusiasts in northern Lincolnshire found themselves longing for the excitement of spotting exotic locomotives amidst the usual daily traffic, and summer weekends brought the thrill of seaside excursion trains, ferrying holidaymakers from Yorkshire to the resort. Despite the lack of real-time train information, enthusiasts eagerly anticipated these excursions, which sometimes brought locomotives from far-flung regions.

We then look at L.N.E.R. wartime mileages and delve into the meticulous research conducted by Simon A.C. Martin on the L.N.E.R. locomotives during the war years, particularly focusing on the Gresley Class J6 0-6-0 locomotives. Martin describes his moment of revelation when he discovered a letter signed by Sir Nigel Gresley, which sparked his interest in delving deeper into Gresley's work. Through extensive archival research, he found the Use of Engine Power document, which provided detailed statistics on the performance of each locomotive class during the war.

Nick Deacon takes us to Colwick in Nottinghamshire, chronicling the evolution from a small village to a vital railway centre, focusing on its transformation due to the arrival of the GNR Derbyshire & Staffordshire Extension lines in 1875. It details the rapid growth of infrastructure, including marshalling yards and a loco depot, propelling Colwick to become one of Europe's largest freight yards.

The Alpine Route is our next offering, Phillip Hellawell looks at the complex Halifax, Bradford and Keighley route, Ian Lamb describes seeing a funny looking engine and Geoff Courtney continues his trainspotting odyssey.

David Cullen delves into the remarkable legacy of the L.N.E.R. Garratt locomotive No. 2395, recounting its illustrious career from its inception to its eventual demise. The narrative vividly describes the locomotive's impressive capabilities, including its massive size, unique wheel arrangement, and innovative design features. Despite encountering challenges such as corrosion and operational limitations, the locomotive played a vital role in hauling heavy loads over challenging terrains, earning it the nickname *The Wath Banker*. Through captivating imagery and detailed descriptions, the text pays homage to the ingenuity and power of this mighty locomotive.

Dave Brennand's piece offers a fascinating insight into the history and rarity of British Railways totem signs, particularly focusing on the Eastern Region. The transition from steam to electric traction marked a significant era in railway history, and the installation of fluorescent lighting along routes such as the Shenfield to Southend-on-Sea Victoria line hastened the demise of these iconic signs.

Roger Geach's account of *Deltic Days at Kings Cross* evokes a vivid picture of the railway scene during the 1970s. The bustling activity at Kings Cross, with powerful locomotives like the Deltic standing on the buffer stops, paints a nostalgic image for railway enthusiasts.

Eastern Times is now available three times a year, and is available on a subscription service. Just sign up using the contact details below to ensure your copy is automatically sent to you every four months. *https://ttpublishing.co.uk/transport-books/*, email *admin@ttpublishing.co.uk*, or call us on *01494 708939*.

PETER SIKES, EDITOR, EASTERN TIMES
email: tteasterntimes@gmail.com

Front cover (and inset right): A service for Sutton-on-Sea stands at Louth station in this undated image. Opened by the GNR on 1st March 1848 it was a principal intermediate station between Boston and Grimsby. The signal box pictured, Louth South, was added to the station in 1887.

On 19th February 1941 the station and extensive goods yard were hit by enemy bombs resulting in the loss of the life of a fireman who was on shunting duties at the time. The station closed to passengers on 4th October 1970 and the was track lifted in 1979/80. Complete closure occurred on 31st December 1980. *Photo: Gerald Daniels © Transport Treasury*

A DAY TO REMEMBER

BY PAUL KING

As Joni Mitchell sings in Big Yellow Taxi, 'You don't know what you got 'til it's gone.' That is very true of the railway scene in the late 1950s and early 1960s. In northern Lincolnshire we had to be content with the mundane mixed traffic and freight locos, the workhorses of the railway system. Spotting the exotic L.N.E.R. Pacifics were adventures we looked forward to, pocket money saved for an all too brief visit to the likes of Retford or Doncaster. All we had were B1s, D11s, J6s, J11s, K2s, K3s, O2s, O4s, WDs, latterly 9Fs and Britannias. For shunting we had J50s, J63s and J94s. Nothing exciting or exotic. I refer you back to the Joni Mitchell quote at the beginning of this article.

24th August 1952 • Class D20 4-4-0 No. 62360 at Cudworth station in the process of coupling onto the Hull & Barnsley Railtour, organised by the Stephenson Locomotive Society/Manchester Locomotive Society standing in the Down slow platform. The loco would work the tour over the Hull & Barnsley Railway to Hull Paragon and back to Cudworth. The three excursions would have reversed along this route onto the H&BR before heading forward to Cleethorpes. *Photo: Boulter-Hancock, Paul King Collection*

However, there was the seaside resort of Cleethorpes and just down the coast Mablethorpe and Skegness. The summer was eagerly looked forward to as then, at weekends, everything changed. Saturdays brought train loads of holidaymakers from the North Midlands and South and West Yorkshire for a week's holiday by the sea. Sundays brought the day trippers. We would stand at Wellowgate in Grimsby waiting for the 4.2 bell in the adjacent signal box that signified an excursion for Cleethorpes. No Realtime Trains or internet, it was all down to observation, occasionally we had lists of excursions due, but never what was hauling them.

Anticipation was intense as we waited to see what would appear under Deansgate Bridge. Sometimes it would be one of our own B1s or K3s, purloined by Lincoln, Doncaster or Sheffield, or one of their own. Sometimes Doncaster would use an ex-works loco, other sheds used whatever was to hand and the locos could come from anywhere on the Eastern or North Eastern regions. Anything from V2s, more accustomed to main line work, to J11s, which were possibly working the pick-up goods the previous day. After nationalisation ex-L.M.S.R Black 5s, Crabs and even 4F 0-6-0s and 8F 2-8-0s were not uncommon as the old company borders began to crumble.

As train spotters we gave little thought to who the occupants of these excursions were. This article is about one day when a small community in South Yorkshire travelled to the east coast and the trip was such that it is part of northern Lincolnshire railway folklore.

The Miner's Welfare was once a systemic part of life in the villages of South and West Yorkshire, in fact they could be found anywhere where coal was mined. Many still survive in the guise of social clubs. As well as providing entertainment and social engagement, they also provided

Having successfully returned the railtour to Cudworth, No. 62360 has left the Hull & Barnsley Railtour stock in the Up slow platform and is reversing onto the Down slow past Cudworth Station South signal box. On the right, Class 3P 4-4-0 No. 40726 is waiting to back down onto the train and return it to its originating point, Manchester Central. Photo: H. C. Casserley, Paul King Collection.

help and aided those in need. They were the social centre of the communities. The annual outing to the seaside was another of their responsibilities and an event looked forward to by the whole community. As the big day approached the numbers would mount and special trains organised, often departing from little more than a wooden platform adjacent to the local mine. As the 1950s advanced and the roads improved, buses and coaches would also be hired. I well remember sitting on Laceby Road in Grimsby watching fleets of Yorkshire Traction, Mexborough & Swinton, East Midlands and other operators' vehicles heading for Cleethorpes. However, I digress, railways remained the main provider throughout the 1950s and into the 1960s.

The village of Cudworth, pronounced Cudderth by the locals, is a few miles north east of Barnsley and like many villages in the area relied heavily on the coal industry for employment. The railway came to Cudworth as early as 1840 when the North Midland Railway opened its line to Leeds from Derby, the station was initially known as Barnsley. In 1885 the Hull & Barnsley Railway terminated at Cudworth where an additional platform was provided as well as a loco shed. The Hull & Barnsley line closed in 1967 and the station followed suit in 1968. The line remained open until 1988 when closure was enforced due to mining subsidence.

In 1958 the population in the area was around 10,000 of which around 1,500 would take the trains organised by Staincross Miner's Welfare to Cleethorpes on Sunday 27th July. The records I have show that these excursions were an annual event and Cleethorpes was the destination every other year. The weather was overcast with a hint of rain and temperatures in the mid-teens. Over on the coast they were forecasting showers with bright periods, but this was in the days before regionalised weather forecasts and the hardy people of Cudworth would try and cater for any weather.

Three trains were booked to take them to Cleethorpes. I have not been able to ascertain the actual start times, but recognising the leisurely pace of excursion trains, timings had to be such as the motive power provided could vary from large semi-express locos down to humble 0-6-0s, more accustomed to earning their keep on goods trains. I have tried to calculate back from the arrival times at Cleethorpes. Therefore, I have estimated the start time as being between 8.30 and 9.30am.

Motive power for the trains was to be provided by Leeds (Holbeck) with the stock coming from Wakefield. The whole of West Yorkshire had become North Eastern Region territory in 1957 but little had changed and Holbeck and Wakefield were still very much ex-L.M.S.R sheds. Had anyone in the Grimsby area seen a list of excursions for that day speculation would have been rife as to what to expect on the three trains from Cudworth. Black 5s or Crabs were favourite but Leeds used any visiting loco so they could be Carlisle or Liverpool based, could one of them be a Jubilee, Holbeck had a lot of them at the time? At this point it is only fair that you understand I have no memory of the event as it was just before my interest began to seriously develop but I have friends who remember it clearly.

One would think it was easy, especially before the mass closures of the 1960s, for trains to run between point A and point B. The empty stock for the three excursions would arrive in Cudworth facing south and one would expect that, once loaded, they would head off in that general direction before turning east for the coast. Not so. After much research a possible route was discovered but it wasn't ideal and would have to use a section of line that was freight only after an excursion bound for Cleethorpes derailed there, this was the ex-Midland Railway curve from Wincobank West to Wincobank North junctions. I appealed for help and eventually, through a colleague in the Great Central Railway Society, was put in touch with someone who described arrangements made for three excursions from Cudworth to Cleethorpes in 1958. The significance of the year evaded me at first but then, on re-reading, I realised he was referring to the subject of this article.

At the time, Cudworth had five platforms, four on through lines, from west to east these were Down Main, Up Main, Down Slow, Up Slow, the fifth, on the eastern side of the station, was the connection to the Hull & Barnsley Railway. To avoid unnecessary manoeuvres, having studied a station plan for Cudworth, I would suggest that the excursions loaded at the platform on the Up slow line, however, if more than one excursion was loading at the same time it would probably have used the Down slow platform too. It is unlikely that all three were loading at the same time as there wasn't a suitable connection between the main lines and the slow lines. The Hull & Barnsley platform appears to have been out of use by this time as it was frequently used for wagon storage.

21st April 1962
The first of the Cudworth excursions was hauled by 45692 *Cyclops* from Perth shed (63A). It is seen here later in its career when allocated to 67A Corkerhill. I haven't been able to confirm the location of the photograph but believe it may be St. Rollox. The loco was withdrawn eight months later.
Photo: Paul King Collection

45597 *Barbados* was the loco in charge of the second Cudworth train. It is seen here at Bangor, North Wales, on an unknown date. The Jubilees tended to move around between sheds, this was certainly true of *Barbados* when first built but by March 1940 it was at Holbeck where it remained until withdrawn in January 1965.
Photo: Paul King Collection

Royston shed, about 3 miles north of Cudworth, was responsible for workings in the area and was designated to provide a pilot for the morning and evening operations required to set these trains on their designated route. Once loaded the pilot would draw them backwards across Cudworth South Junction (note this was north of the station) to join the spur from the Hull & Barnsley line. The reversal would continue as far as Cudworth Goods Junction where reversal would again take place with the pilot uncoupling and the train engine taking the excursion forward to Stairfoot. At Stairfoot the excursions joined the MS&LR line from Barnsley to Mexborough via Wombwell and Wath. At Hexthorpe Junction they would take the Doncaster avoiding line and cross the ECML just north of Doncaster station. At Bentley Junction they would join the line from Doncaster and head towards the coast through Stainforth & Hatfield before turning east at Thorne Junction. Through Crowle and Scunthorpe to Wrawby Junction and on through Barnetby and Brocklesby until they had to slow for the curve through Grimsby Town station.

It was only then that the young trainspotters at Wellowgate got their first glimpse of the approaching trains. The first excursion through that day was around 10.30 from Lincoln with B1 No. 61202, not uncommon in the area having been at Immingham until 1953 and Lincoln ever since. Five minutes later and it was Doncaster based B1 No. 61193 with a train from its hometown. Next, around 10.45 was Lincoln based K3 No. 61944 with an excursion from Sheffield followed almost immediately by a similar train from Retford headed by No. 61208. Stainforth & Hatfield provided the next train, just after 11 o'clock, hauled by K3 No. 61819 from Hull Dairycoates. A quarter of an hour passed before the next excursion came through, again from Stainforth but B1 hauled, No. 61120 being a Doncaster engine. So far six excursions with nothing exceptional hauling them, even Hull locos were not uncommon as they were often to be found on Doncaster shed and would be used ad hoc for excursion work.

Shortly before 11.45, the bells rang in the signal box, the gates swung across the road, the points were set, signals pulled off. Under Deansgate Bridge came an L.M.S. loco, a green L.M.S. loco. "Jube" was the excited reaction and No. 45692 *Cyclops* majestically steamed into town. Imagine the reaction when the more observant noticed the shedplate, 63A, Perth. A cop for virtually everyone in the area.

Five minutes later the gates closed again, everyone now on tenterhooks as the sight of a Jubilee in the area, although not unique, was rare enough to get the youngsters excited. It was a B1, No. 61123, on an excursion from Hemsworth. Although Ardsley based, locos such as 61123 were often seen in the area on excursion work. Soon after, 4.2 rang out again and the gates closed, the signals were pulled off, the road was set. Some excursions were routed through platform 1 whilst others took the avoiding line around the back of platform 3, no one knew which route would be set. Steam was seen above Deansgate Bridge and through its portal came another L.M.S. loco, another green L.M.S. loco. The second excursion from Cudworth had arrived with Leeds Holbeck's No. 45597 *Barbados*, not necessarily a cop this time as Leeds Jubilees could be seen at Doncaster, Sheffield and York, all venues within easy reach.

Almost 10 minutes passed before the next excursion arrived, a second one from Hemsworth, and the second to have an Ardsley based B1 in charge. This would have caused slightly more excitement as it was a namer, No. 61011 *Waterbuck*. The next to arrive, five minutes behind 61011, was another K3 hauled excursion from Sheffield. No. 61951 would have caused a bit of a stir as the more observant would have noted it carried a 30F Parkeston Quay plate on its smokebox door. There was one more Cudworth excursion on its way and it was the next one signalled to pass. Another L.M.S. loco at the head, another green L.M.S. loco at the head, another Jubilee. No. 45699 *Galatea* carrying an 82E shedplate, Bristol Barrow Road. Almost certainly another cop for everyone in the area. Leeds Holbeck had not only done the folk of

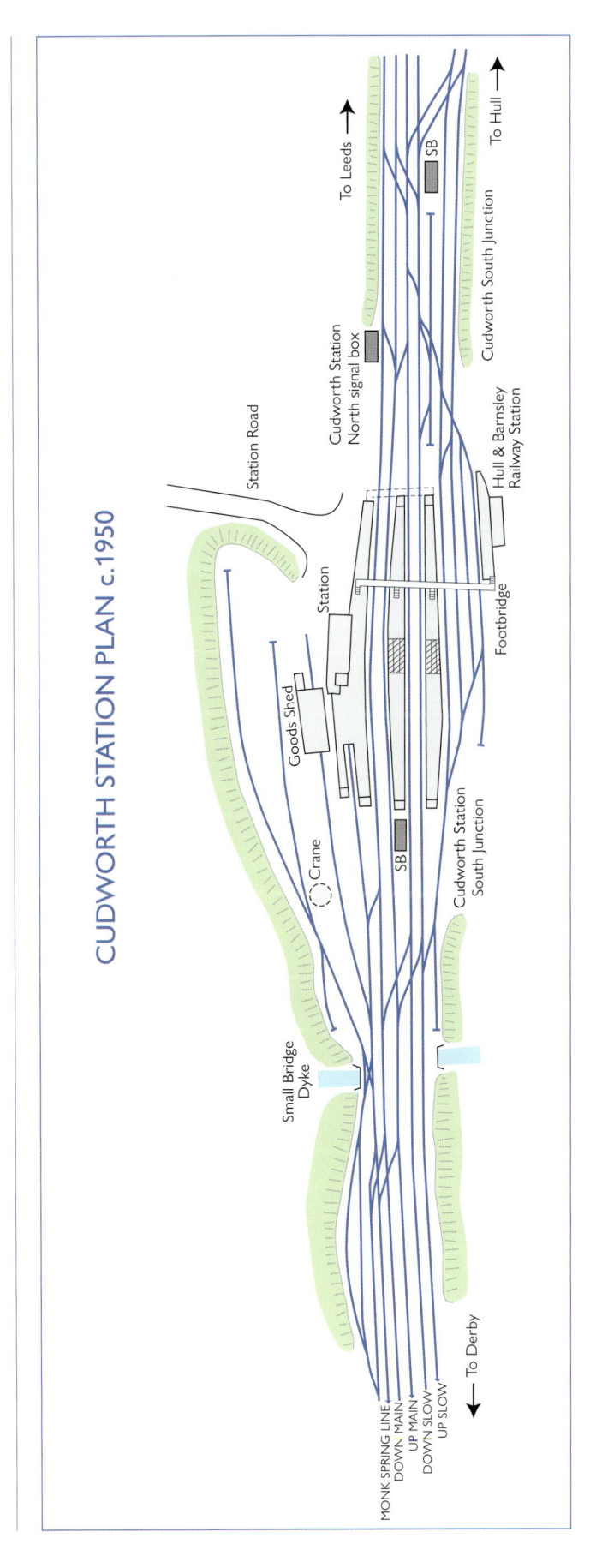

CUDWORTH STATION PLAN c.1950

16th August 1953 • The final Cudworth excursion was entrusted to 45699 *Galatea*. It is seen passing Dore & Totley station with an express for the south-west. Withdrawn in November 1964 it eventually found its way to Woodhams at Barry. Leaving in 1980 it was in a sorry state and was to be used as a source of spares for classmate 45690 *Leander*. Fortunately, that didn't happen and it is now a respected member of the preservation movement. *Photo: Boulter-Hancock, Paul King Collection*

Staincross and Cudworth proud but provided a talking point that still echoes around northern Lincolnshire today. Three Jubilees to Cleethorpes in one day.

A good friend of mine had just started spotting and I believe he was at Suggitts Lane on this day. He remembers the green locos and noting the numbers down but was a bit puzzled as his Ian Allan ABC started at 60001, there were no 40000s. He had the Eastern Region ABC and until then didn't realise there were different numbering systems for each region and different books covering them. As will be seen from the table on page 14 that wasn't the end of the day's entertainment and excursions continued to arrive at Cleethorpes until 3.25 in the afternoon. None of the locomotives were as outstanding as the three Jubilees, although Crab 2-6-0 No. 42704 on the Radford (Nottingham) excursion heralded from Newton Heath shed in Manchester. The Black 5, No. 44694, was from the Bradford shed at Low Moor. Of the B1s, No. 61386 was Copley Hill based and No. 61283 was Cambridge based, the others were either Doncaster or Mexborough allocated and relatively common in the area. The K2, No. 61723, was Colwick based, as one would expect with an excursion from Teversall in the Nottinghamshire coalfield. The two K3s were also locally based, No. 61940 from Doncaster and No. 61839 Immingham. The final loco, B16 No. 61444, like so many of its classmates, was York based and not uncommon, working in on the daily goods from that city.

A total of 26 trains, in addition to the regular services, put tremendous pressure on the operating department and they were a lucrative benefit to the local drivers and firemen. On arrival at Cleethorpes a local crew would take over from the train crew. It was vital that the platforms be vacated as quickly as possible, there were six in use at Cleethorpes at the time, platforms 1 and 2 were, generally, reserved for regular traffic, although at times of extreme pressure they could be used too, whilst platforms 3 to 6 would handle the excursions. The train would reverse out of the station and the loco would uncouple, pull forward and reverse onto the turntable behind the signal box. Whilst in the platform or at this point the fireman would pull forward the coal in the tender, occasionally the loco would have to be sent to Grimsby loco for coal. Once turned and serviced, the loco would couple onto what had been the rear of its train and take it to New Clee Sidings, about a mile away, for storage. The crew would then walk back to the station and do the task again, and again. The number of times was dependent on the number of crews on duty. As evening approached these same local crews would return to New Clee and reverse the excursions back into the platforms at Cleethorpes, repeating the process until all the trains had been dealt with. Only then could they go home and put their feet up before resuming their ordinary duties the next day. It was a hard day's work for these men, but the benefits, financially, made it worthwhile.

I have records for the period from June 1958 through to September 1962. Most Sundays there would be at least 20 excursions arriving at Cleethorpes, often nearer to 30. The record was in 1956 when more than 50 arrived in one day. The most common power were B1s and K3s. All 10 of the Class D11 Improved Directors were regular visitors during the latter years of their career as were the ex-GNR K2 2-6-0s. The occasional V2 usually appeared from Doncaster whilst B16s often brought in excursions from the southern reaches of North Yorkshire. The strangest was an excursion from Thirsk in North Yorkshire on the 19th July 1959, which arrived behind B17 No. 61641 *Gayton Hall*, a March loco that was withdrawn the following January.

In the late 1950s Black 5s were also popular, particularly on trains from West Yorkshire and Nottinghamshire. Two years after the Jubilees, 17th July 1960, the same set of excursions were worked from Cudworth by three members of the class, No. 44666 based at Saltley, No. 44667 at Leicester Midland and No. 45082 which was a Carlisle Kingmoor loco. On this occasion 45082 and 44666 arrived within four minutes of each other at 9.55 and 9.59, the third didn't arrive until 11.12. One must consider that 44667 was struggling with its train as it is noted that it was replaced by New England based B1 No. 61060 for the return journey. These were three of twenty-two excursions that day. Thirteen were hauled by B1 4-6-0s, three by K3s, an excursion from Radford (Nottingham) brought Crab 2-6-0 No. 42726 and the final arrival of the day produced Standard 5 4-6-0 No. 73002. An excursion from Sheffield has the loco described as DMU with no further information.

It would seem the person writing the log in Cleethorpes signal box was as interested in these vehicles as we spotters were at the time.

The visit to Cleethorpes appears to have been a biennial event. Two years later three excursions duly arrived from Cudworth on 22nd July 1962. By now, diesels were beginning to take a hold and of the twenty excursions on that day, no less than ten were in the hands of Sheffield Darnall based Brush Type 2s and two were DMUs. The remaining eight were entrusted to two Black 5s, two K3s and four B1s. The Cudworth ones were in the charge of three of the Brush Type 2s, D5686, D5690 and D5806.

The furthest flung of the Black 5 visitors was No. 45453 on 2nd August 1959 with an excursion from Nottingham Midland although the loco was from Perth. There was also a smattering of the BR Standard Class 5 4-6-0s, including three of the Caprotti valve gear ones. Other L.M.S. types were the mixed traffic Crab 2-6-0s, there were also three Stanier 2-6-0s recorded in the five years I have records for, six Class 8F 2-8-0s were also recorded along with a couple of 4F 0-6-0s. As for Jubilees, beside the three in this article the records show No. 45562 *Alberta* on 21st May 1961 from Leeds and Nuneaton based No. 45624 *St. Helena* from Moira, in north west Leicestershire close to the Derbyshire border. From memory, *Alberta* visited more than once, as did other Holbeck based classmates, there is photographic evidence of No. 45589 *Gwalior* at Suggitts Lane, No. 45598 *Basutoland* on 21st June 1963 leaving Cleethorpes and No. 45675 *Hardy* at Grimsby Town, there were others but the dates have been lost in time.

Returning to the residents of Staincross and Cudworth, what could they expect to see whilst at Cleethorpes? As well as the delights of Wonderland and the North Promenade there was the Big Dips, as it was known with the miniature railway weaving its way amongst the girders of this roller coaster ride, deckchairs and donkeys on the beach, fish and chips and ice cream. A few probably partook of Hewitt's Brown Jack or other, equally, enjoyable ales in the local establishments. There was the open-air bathing pool, whose freezing waters were enjoyed by many a holidaymaker and local alike over the years. Nearby the Café Dansant was offering the Winter Gardens Salon Orchestra playing a selection of light music in the Palm Court.

Looking from the promenade at the ships in the Humber Estuary the visitors would see the trawlers waiting in line to enter the fish dock at Grimsby on the afternoon tide, which was 4 o'clock on 28th July. The tug *Brenda Fisher* would be fussing around the ships, making sure they knew the order to enter port. Amongst the vessels waiting to land their catch for the Monday morning fish market were the Grimsby based trawlers *Kelly*, *Joseph Knibb*, *Ross Tiger*, *Hull City*, *Alsey* and *Port Vale*, seine netters *Ebor Jewel* and *Scanlord*, plus the Belgian registered *Charvic* and *Belgian Sailor* and the Danish seine netters *Corneliussen* and *Dakota*.

If they had been staying for more than the day, the evening show at the Empire Theatre on Alexandra Road offered Bobbie Thompson, Northern Comedian, The Liddell Triplets (television's tuneful triplets), Victor Seaforth (famous vocal comedian), Johnny and Suma Lamonte, The Skating Merenos, Sunny Rogers, The Goldwyns, Johnny Maxim, in their words – Another Tip-Top Variety Bill with shows at 6.15 and 8.30. The Theatre Royal, at the top of the station approach, was showing *Don't Go Near The Water* starring Glenn Ford, Keenan Wynn and Russ Tamblyn with *Sierra Stranger* as the support film. Down Grimsby Road, adjacent to Grimsby Town's football ground, the Ritz cinema was showing *A Night To Remember* starring Kenneth More, Anthony Bushell and Honor Blackman with *Gaucho Country* as the support film. '*A Night To Remember*', well, those 1,500 residents of South Yorkshire had certainly given the trainspotters of northern Lincolnshire a day to remember.

Note: This article has been in preparation for more than five years and was intended to complement a chapter in part 4 of my series of books. However, my research on the earlier part of the journey drew a blank due to being unable to show how the trains involved could reach their destination without reversal and a great deal of tender first running, not a common occurrence with large engines on passenger workings.

28th June 1959 • No book referencing excursion traffic and Cleethorpes would be complete without this view of New Clee carriage sidings. Nine excursions await the summons back into Cleethorpes and they are, from left to right: B1s Nos. 61165 (Mexborough), 61377 (Doncaster), 61230 (Hammerton Street), D11 No. 62668 *Jutland* (Darnall), K3 No. 61824 (Woodford Halse), B1s Nos. 61208 and 61231 (both Retford), K3 No. 61803 (Doncaster) and D11, now preserved, No. 62660 *Butler Henderson* (Darnall). Their destinations, respectively, were Rotherham, Doncaster, Fitzwilliam, Sutton-in-Ashfield, Sheffield, Worksop x2, Harworth Colliery and Kirkby-in-Ashfield.
Photo: Neville Stead © Transport Treasury

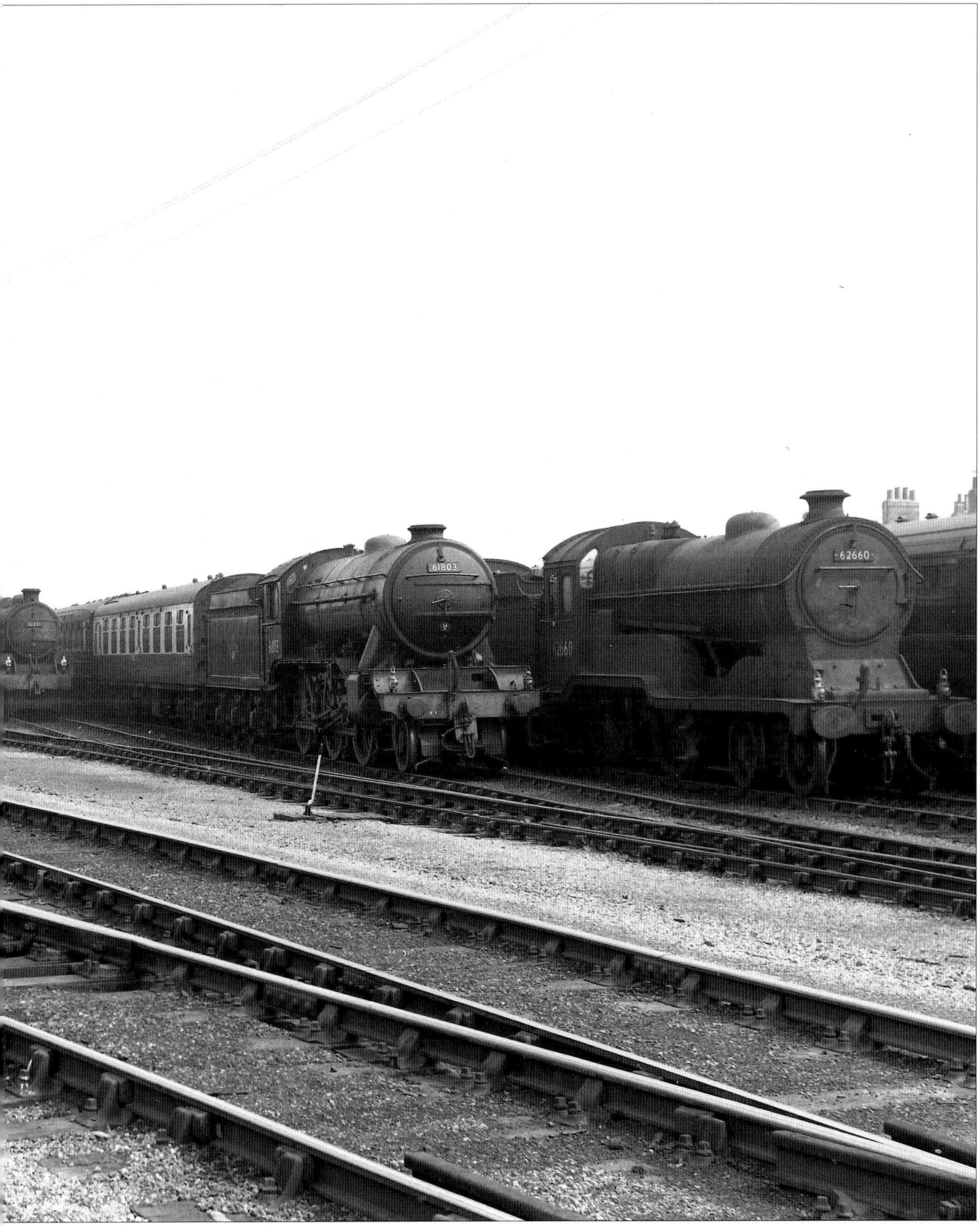
61803
62660

The tables below show the excursions arriving at Cleethorpes on the three dates of the Cudworth trains described in the text. Alongside the originating point is their departure time.

SUNDAY 27th JULY 1958

Arrival	Loco	Originating Point/Return	Arrival	Loco	Originating Point/Return	Arrival	Loco	Originating Point/Return
10.48	61202	Lincoln/16.30	12.13	61011	Hemsworth/18.34	14.05	61377	Doncaster/18.48
10.55	61193	Doncaster/16.00	12.19	61951	Sheffield/18.12	14.05	61165	Rotherham/19.05
11.00	61944	Sheffield/16.15	12.25	45699	Cudworth/18.05	14.11	44694	Bradford/19.23
11.04	61208	Retford/17.24	12.57	61114	Scunthorpe/19.51	14.27	61723	Teversall/18.27
11.20	61819	Stainforth/16.33	13.10	61283	Sheffield/18.27	14.33	61162	Barnsley/19.44
11.35	61120	Stainforth/16.43	13.15	61940	Barnby Dun/15.22	14.40	61386	Leeds/19.09
11.55	45692	Cudworth/17.30	13.15	61839	Deepcar/18.55	15.05	61444	Ferrybridge/21.05
11.59	61123	Hemsworth/17.44	13.21	42704	Radford/17.58	15.25	61285	Penistone/20.27
12.05	45597	Cudworth/17.51	13.35	61167	Mexborough/20.00			

SUNDAY 17th JULY 1960

Arrival	Loco	Originating Point/Return	Arrival	Loco	Originating Point/Return	Arrival	Loco	Originating Point/Return
09.55	45082	Cudworth/16.32	12.05	61042	Kiveton Bridge/20.09	13.27	61331	Teversall/17.30
09.59	44666	Cudworth/16.57	12.10	61276	Castleford/16.18	13.51	61157	Conisborough/19.02
10.30	61831	Langwith/16.03	12.17	61026	Lincoln/18.41	14.05	61125	Penistone/17.58
10.40	61921	Doncaster/15.56	12.25	61152	Castleford/17.33	14.12	61193	Rotherham/18.12
11.05	DMU	Sheffield/16.25	12.59	61804	Sheffield/18.34	14.35	61182	Bawtry/19.50
11.12	44667	Cudworth/17.30	13.06	42726	Radford/17.44	14.45	61339	Leeds/17.51
11.27	61127	Finningley/18.05	13.13	61196	Deepcar/18.49	15.23	73002	Wombwell/19.35
11.59	61194	Kilnhurst/15.48	Note: 44667 replaced by 61060 and 61276 replaced by 61325 (assumed loco failures)					

SUNDAY 22nd JULY 1962

Arrival	Loco	Originating Point/Return	Arrival	Loco	Originating Point/Return	Arrival	Loco	Originating Point/Return
08.55	61942	Thorne/19.25	11.02	61040	South Elmsall/15.48	13.21	D5690	Cudworth/17.36
09.55	45290	Castleford/17.30	11.47	D5817	Woodhouse/15.48	13.45	D5806	Cudworth/18.12
10.10	44856	Bulwell/15.41	•	61374	New Holland/•	14.02	D5687	Kiveton Bridge/18.40
10.15	D5831	Bolton-on-Dearne/17.40	12.30	D5809	Conisborough/16.17	•	61145	Doncaster/•
10.23	61161	Edlington/16.18	12.53	D5847	Conisborough/16.39	14.28	D5806	Penistone/18.47
10.39	61972	Doncaster/15.16	13.00	D5834	Chesterfield/17.43	14.40	DMU	Bradford/18.42
11.02	DMU	Sheffield/16.40	13.07	D5686	Cudworth/17.08	• Denotes no time shown		

One of the delights awaiting the visitors from Cudworth were the Big Dips, part of the structure is visible on the left, and the miniature railway which ran around it. There was also a boating pool, hardly a lake, in the middle with motor boats for hire. The water was that shallow that too many people in one boat and it would sit on the bottom and refuse to move. One of the regular locos in use, *Henrietta*, is seen rounding the north end of the line and heading towards the station at Wonderland. In the background are the carriage sidings at Suggitts Lane, New Clee sidings were further to the west. *Henrietta*, after many years roaming the country, is now back in the area and in regular use at Waltham Windmill. *Photo: Paul King Collection*

A scene from the early years of the 20th century, it has been impossible to date it exactly. This view of the beach at Cleethorpes shows how crowded the resort was, it remained so into the 1950s and '60s with barely a speck of sand visible. The refreshment rooms and the clock tower at the station are visible on the left.
Photo:
Edward Trevitt Collection, courtesy Wendy Trevitt

17th May 1959 • The final excursion of the day into Cleethorpes brought ex-works Hull Dairycoates based Class K3 2-6-0 No. 61897 with an excursion from Newcastle. It didn't arrive until 3.30pm and left at 8.45pm in the evening. The train is seen here approaching New Clee sidings after servicing and is opposite Grimsby Town's football ground with Fuller Street footbridge in the background. This loco would not have hauled the train from Newcastle, it probably relieved a Gateshead engine at Doncaster.
Photo: Neville Stead @ Transport Treasury

17th July 1960 • Unfortunately, not of the best quality, this photograph captures one of the locos arriving with an excursion on the day the three Black 5s came with the Cudworth trains. B1 No. 61026 *Ourebi* is seen in platform 5 after bringing in an excursion from Lincoln. 61026 was a familiar sight in the area and was affectionately known by a play on its name, *Ourebi* becoming 'Our B1'. New to Doncaster in 1948, it moved to Lincoln in 1957. Immingham was home three times in between a further spell at Lincoln and two months at Colwick in late 1965. Returning, for the last time, in December 1965 it was withdrawn from Immingham in February 1966 when the shed closed to steam.
Photo: Paul King Collection

15th March 1961 • Immingham B1s had a hard life, being expected to handle the Cleethorpes–London trains prior to being superseded by the Britannias in 1961. They were also expected to work to Birmingham, and occasionally beyond, Manchester and Leeds. More mundane duties also came under their portfolio and on the 22nd July 1962, 61374 handled an excursion from New Holland. It is seen here just over a year earlier passing Belle Isle with one of the morning expresses from Cleethorpes. Note the larger Goods & Mineral signal box, overseeing the movements from and to Top Shed on the banking on the left and the smaller Belle Isle signal box behind the signal gantry. 61374 spent its entire working life based at Immingham, albeit a mere 12½ years from February 1951 to September 1963. *Photo: R. C. Riley © Transport Treasury*

L.N.E.R. LOCOMOTIVE MILEAGES AND AVAILABILITY IN THE WAR YEARS

BY SIMON A.C. MARTIN

PART 1: THE USE OF ENGINE POWER AND THE GRESLEY J6s

The best moments in reading are when you come across something – a thought, a feeling, a way of looking at things – which you had thought special and particular to you. And now, here it is, set down by someone else, a person you have never met, someone even who is long dead. And it is as if a hand has come out and taken yours.

Alan Bennett, English playwright.

There was a moment where I felt a hand had reached out and taken mine when I sat in the National Archives at Kew. It was mid-afternoon in 2018, during the midst of my research into Edward Thompson, when I found a letter signed by Sir Nigel Gresley for the first time.

The letter was to the point and kindly written, asking Thompson to arrange for the fitting of the Kylchap chimney to the water tube boiler locomotive, the W1. Gresley thanked Thompson for his support in the matter of the modifications; Thompson was of course now the head of Darlington Works.

There was a warmth to Gresley's requests and in Thompson's responses that I couldn't quite shake from my mind. Was this me projecting an idea onto this evidence? Subjectivity exists in every interpretation of the primary evidence.

I felt determined after completing my tome on Thompson that I should return to Gresley and follow up with a book I had always wanted to write.

The more I delved back into the archives, looking through letters, reports, drawings, graphs, board minutes and the rest, the more convinced I was that the issues between the two men had been played up and exaggerated greatly over the years. The substance of the communications between them was always couched in professional terms.

When I discovered the Use of Engine Power document (a report which gave average annual mileages and availability statistics for every locomotive class operating on the railway) I was astonished at the amount of data that was available to analyse for the L.N.E.R. That this document survived to the modern day is remarkable. What was perhaps more remarkable is that until my tome on Thompson in 2021 no one else had ever acknowledged its existence in print.

Where Edward Thompson was concerned it was easier to step back and let the figures do the bulk of the talking. The Use of Engine Power document was explicit in showcasing the excellence of his design work during the Second World War and nothing more was required to be said.

Where Gresley is concerned the very same document shows us a different side to his work. This is a very good thing. Far from what some would anticipate would be an undoing of Gresley's reputation, my findings are likely to further cement that well-deserved reputation. There is no doubt that he provided the G.N.R. and then L.N.E.R. ample new motive power that covered the requirements of these railways' needs.

There is also little doubt that Gresley innovated beyond all other locomotive engineers at every aspect of the railway including locomotive and Carriage & Wagon design, signalling, track, electrification and many more besides.

It is not unfair to state that the modern East Coast Main Line (E.C.M.L.) owes much to Gresley's work throughout his life. Without Gresley proving the clear advantages of streamlining an entire train, the benefits of articulation, the need for a bigger picture thinking for high-speed operations, it is doubtful that the United Kingdom would have been at the forefront of many of the technical innovations that have led to the modern railway.

What we see being designed for High Speed 2 (HS2) now in terms of infrastructure, rolling stock and signalling equipment are all echoes of that Gresley was pushing for in the 1930s. Gresley gave the world its first truly high-speed railway. Nothing before and nothing since has made the same level of impact that *Silver Link* did in September 1935.

Notes on how the data is presented and interpreted

At the end of the locomotive section within this article there will be a summarisation of the statistics taken from the London & North Eastern Railway's internal Use of Engine Power document. These statistics are in relation to the individual classes' average annual mileages and availability.

This document, available in the National Archives at Kew, is a record of statistics on every locomotive class that was owned or operated by the L.N.E.R. during the Second World War. These statistics record the following data:

Number of days off for repairs

At Shops
Stopped and waiting to go to Works
Sent to Works and in Works

At Sheds
Under depot repairs
Under depot repairs – awaiting material
Under examination at depots
Waiting depot repairs
Waiting repair decision at depots
Stopped for boiler washing

Number of days engines available

In use
In use at home and other depots

Not in use
Available but not in use at home and other depots
Engines on loan to government, etc.
Engines tallowed down

The Locomotive Accountant's Office
STRATFORD E.15

Fig.1.
How data was recorded in the Use of Engine Power Document.

These statistics were collated centrally from the Locomotive Accountant's Office in Stratford originally. The associated papers attached to this part of the L.N.E.R.'s archive suggest strongly that Sir Nigel Gresley originally initiated the production of this document to have an overview of and to investigate more fully the availability issues of the L.N.E.R.'s locomotive stock in the run up to the Second World War.

On examination of the original document, I took the decision to collate the entire database into a modern format, in the form of a spreadsheet. This process took around three years with much discussion amongst peers on how to fully analyse the data. It was realised early on that the dataset the document held allowed, potentially, a closer view of the day-to-day performance of the locomotives being used. The solution was the production of a tailored Availability formula, as outlined below:

% availability has been calculated by using the following formula:

(Number of days engines in a class are available) divided by (Number of engines in a class) x (Number of work days in a year – England and Scotland)

Fig. 2.
How availability of a locomotive class in a year is calculated.

This formula adjusts for the different number of days available for work in England and Scotland, within the database I produced.

For the purposes of the consistency of the database the original figures are maintained with appropriate explanations in the notes. This leaves us with a modern version of the Use of Engine Power document.

The L.N.E.R. split these statistics into four basic areas of the railway, as outlined below (including their letter codes):

SAW	Southern Area Western Section
SAE	Southern Area Eastern Section
NEA	North Eastern Area
SCO	Scottish Area
WHL	Whole Line

Fig. 3.
How availability of a locomotive class in a year is calculated.

The dataset we have is huge. It covers between 6,200 to 6,500 steam locomotives in any given year, sub-divided into around 160 classes, covering four major areas of the

Pictured at an unrecorded location, ex-GNR Ivatt Class C1 4-4-2 No. 4423 looks majestic after being prepared to haul the 'Queen of Scots', a Pullman service that ran from London Kings Cross to Glasgow Queen Street via Leeds Central, Harrogate, Newcastle and Edinburgh Waverley.

No. 4423 entered service in April 1907 as No. 1423, being renumbered 4423 by the L.N.E.R. in October 1924 and then again to 2853 in November 1946. The loco was withdrawn from service at Doncaster Works in May 1947 after a service life of just over forty years.

L.N.E.R. with a further set of statistics covering the rest of the L.N.E.R. That gives us around 60,000 individual data entries over the five years recorded.

The Use of Engine Power document tells us far more about the clear day to day workings of the railway than any simplified set of works visits or withdrawal dates. For the first time, we can see the real picture of the L.N.E.R.'s locomotive stock in work.

For the purposes of this book, only the designs directly attributable to Gresley have been included. For further information on later designs from Edward Thompson please refer to *Edward Thompson: Wartime C.M.E.*, that is also available from Strathwood Publishing Limited.

The examination of each classes' individual performances by way of the Use of Engine Power document is specific to the war years and should be considered in that context. It may inform how they performed in peacetime to some degree, and for each locomotive class some supposition may be observed. Where we have more data, such as that from the Engine Record Cards held at the National Railway Museum we can use this to contrast with, or emphasise further, the Use of Engine Power data.

For the purposes of analysing the data that we have, the Whole Line Averages will be used to illustrate the relative availability and mileages of any given class across the whole of the L.N.E.R.

Totals	1942	1943	1944	1945	1946
Total number of locomotives	6,237	6,401	6,430	6,273	6,451
Total number of locomotive classes	164	163	160	162	162
Total number of boiler types	193	199	173	172	172
Average availability for whole fleet	71%	71%	73%	71%	68%
Total fleet mileage	151,829,377	155,311,171	151,782,938	155,377,464	156,619,241
Difference in number of locos	n/a	164	29	-157	178
Difference in number of classes	n/a	-1	-3	2	0

Fig. 4. An overview of the war years 1942–44. This high-level overview of the whole L.N.E.R. locomotive fleet gives an indication of the railway's issues and successes.

L.N.E.R. Class A4 No. 14 *Silver Link*. Photo: J. T. Rutherford © Transport Treasury

GRESLEY CLASS J6 0-6-0 (1911)

NUMBER BUILT (G.N.R.) – 110

Strictly speaking, the class J6 was the final development of the Ivatt line of G.N.R. 0-6-0 locomotives, however by L.N.E.R. days it was recognised that the changes made by Gresley with the superheated versions were significant enough to justify being classified differently to the earlier Ivatt locomotives.

A total of 95 J6s were built under Gresley's direction, all at Doncaster works in ten batches over the course of a decade with the original Ivatt-era locomotives being added to the Gresley development to create a total class of 110 locomotives.

All the J6s were initially built with standard G.N.R. Ramsbottom safety valves. As boilers came up for repair in the L.N.E.R. era, the J6s were gradually converted to being fitted with the L.N.E.R.'s chosen standard, Ross Pop safety valves.

There were various experiments with superheaters on the J6s with the first few batches being built with Schmidt superheaters with 18 elements. There were also five locomotives which were fitted with Robinson superheaters, and six fitted with a type of superheater colloquially known as the 'Doncaster Straight Tube' type. This type differed from the others in that it had vertical headers which held the elements in place. In the boilers fitted with this arrangement there were separate saturated and superheated headers.

The fourth type of superheater was applied to one locomotive only in 1913 (No. 563) which was a Gresley design known as a 'Twin Tube' superheater. This was then replaced by a 'Triple Tube' superheater two years later. Improvements included an increase in superheated surface area and had 17 elements.

The trials gave a mixture of results, and by the early 1920s the class had the choice of superheater reduced to two types; the Schmidt type (fitted to nine locomotives) and the other 101 locomotives used the Robinson design. By 1927 the L.N.E.R. had fully standardised on the Robinson design of superheater across their locomotive fleet and the remaining J6s with Schmidt type superheaters were converted accordingly.

Gresley under the G.N.R. experimented with feed water heating and several types provided by different companies between 1916 and 1918. These experiments were followed up with feed water equipment being fitted to other Gresley locomotives over the next two decades on the L.N.E.R., including classes A3, B12, O2 and P2. The feed water heaters ranged from Dabeg to Auxiliaires des Chemins de Fer (A.C.F.I.) equipment and without exception were all removed after short periods of time in service. Three members of class J6 received such equipment in 1928, with the feed water heaters being removed by the end of 1932.

The J6s were initially intended as fast mixed traffic locomotives for pulling express goods trains, but by the grouping in 1923 had been cascaded down onto coal trains and local goods trains.

They were mostly concentrated in the western area of the southern section of the L.N.E.R. with a large allocation at Colwick over the years (between 35 and 40). A good number were also allocated at Doncaster and New England with the rest spread sporadically across the L.N.E.R., including a pair kept in London for some time too. There were seven members of the class which were sent to cover a group of overhauled J25s that had been loaned to the Great Western Railway (G.W.R.) in 1940 but with the return of the latter engines they were sent back to their usual haunts. The Nottinghamshire based J6s were only displaced from passenger services with the arrival of brand-new Thompson L1 tank locomotives in the early 1950s.

WARTIME MILEAGES AND AVAILABILITY

The mileages and availability of the Gresley J6s throughout the Second World War has been recorded in the Use of Engine Power document. They achieved respectable availability, with little change year on year and remaining in and around the 80% availability mark. Their average mileages peaked in 1942, falling in 1943, with a small improvement in 1944 but for the next two years continued to fall. This was in line with the average fleet availability and with only a few thousand miles on average a year less than those a few years earlier.

Year	Mileages	Availability
1942	24,475	81%
1943	22,411	79%
1944	23,331	85%
1945	22,525	81%
1946	21,544	80%

Fig. 5. J6 Mileages and availability statistics.

The J6s were not included in Thompson's standardisation scheme and were also absent from L.N.E.R. literature through to British Railways days. Although all the Gresley J6s survived the Second World War, by British Railways days they were beginning to show their age and the first withdrawals started in 1955. By the end of 1962, the entire class had been scrapped.

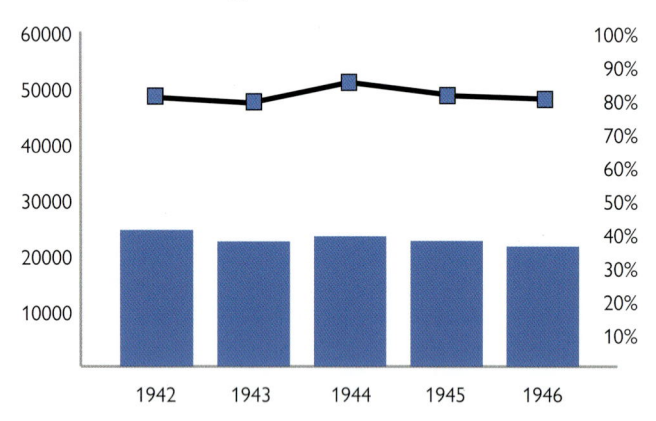

Fig. 6. J6 Mileages and availability graph.

Bibliography

All table and graph figures:
Use of Engine Power document by Simon A.C. Martin. Originally published in 'Sir Nigel Gresley: The L.N.E.R's first C.M.E.', available from Strathwood Publishing Limited.

Publications:
'Edward Thompson of the L.N.E.R.', Peter Grafton (1971 and 2007) Kestrel Books and Oakwood Press.
'Edward Thompson: Wartime C.M.E.', Simon A.C. Martin (2021) Strathwood Publishing Ltd.
'FORWARD, The L.N.E.R. Development Programme', L.N.E.R. (1946) Waterlow & Sons Limited.
'Gresley's Legacy: Locomotives and Rolling Stock', David McIntosh (2015) Ian Allan Publishing Ltd.
'Gresley Locomotives', Brian Haresnape (1981) Ian Allan.
'Gresley and his Locomotives', Tim Hillier-Graves (2019) Pen & Sword Books Ltd.
'Locomotives of the L.N.E.R.', Volumes 1-11, The Railway Correspondence and Travel Society (R.C.T.S.) (1963-1971).

Engineering Papers and Journals:
'The Development of L.N.E.R. Locomotive Design, 1923-1941', Bert Spencer (1941) in an address to the IMechE.

National Archive Materials:
The L.N.E.R. Board, Emergency Board and Locomotive Committee Minutes: File RAIL 390.
The L.N.E.R. Assorted Archives (1923-1948): File RAIL 394.

National Railway Museum Materials:
L.N.E.R. Pacifics Engine Record Cards.
The Edward Thompson Archive.

Below: **11th April 1935 • L.N.E.R. J6 0-6-0 No. 3576 at Nottingham Victoria.** *Photo: George Barlow © Transport Treasury*

L.N.E.R. Class J6 No. 3589 on an Up freight passes Trent signal box at North Muskham on 28th April 1938. *Photo: George Barlow © Transport Treasury*

Class J6 No. 64172 pictured at Peterborough in 1951. *Photo: Jim Flint/Jim Harbart © Transport Treasury*

Gresley J6 0-6-0 No. 64244 peeps out from Boston shed on 13th March 1955. *Photo: Eric Sawford © Transport Treasury*

COLWICK – A FORGOTTEN GREAT NORTHERN EPICENTRE

BY NICK DEACON

EARLY YEARS

An observant reader of the 1938 Nottinghamshire edition of *The King's England* would have noted that the modest inclusion of Colwick (pronounced locally as 'Collick') did not mention anything of the large concentration of railway yards known by that name. The same reader may have also noted that Netherfield, the actual location of these yards to the south-east of Nottingham, did not deserve a single mention in the book probably because as a minor entity of Colwick it did not possess a similar weight of historical importance as its neighbour.

A superb 1932 portrait of one of the many Robinson-designed O4 2-8-0s which arrived at Colwick during the 1920s. This is ex-ROD No. 6624 built by the North British Loco Co. in August 1919 and purchased by the L.N.E.R. in August 1928 and classed as O4/3 because of the lack of tender scoop and having only a steam brake. With the Thompson regime calling for a standard heavy goods loco the numerous O4 series was deemed as being appropriate for development and rebuilds incorporating the standard 100A boiler, standard cylinders and Walschaerts valve gear were taken in hand from February 1944. Colwick-based No. 6624 was one such conversion/rebuild completed at Gorton between July and September 1944 and the loco reappeared transformed and reclassified as Class O1. In 1946 it was renumbered to 3867 and in 1949 (as BR No. 63867) it was then based at Annesley where it remained until withdrawn in November 1962. Photo: David P. Williams Colour Archive.

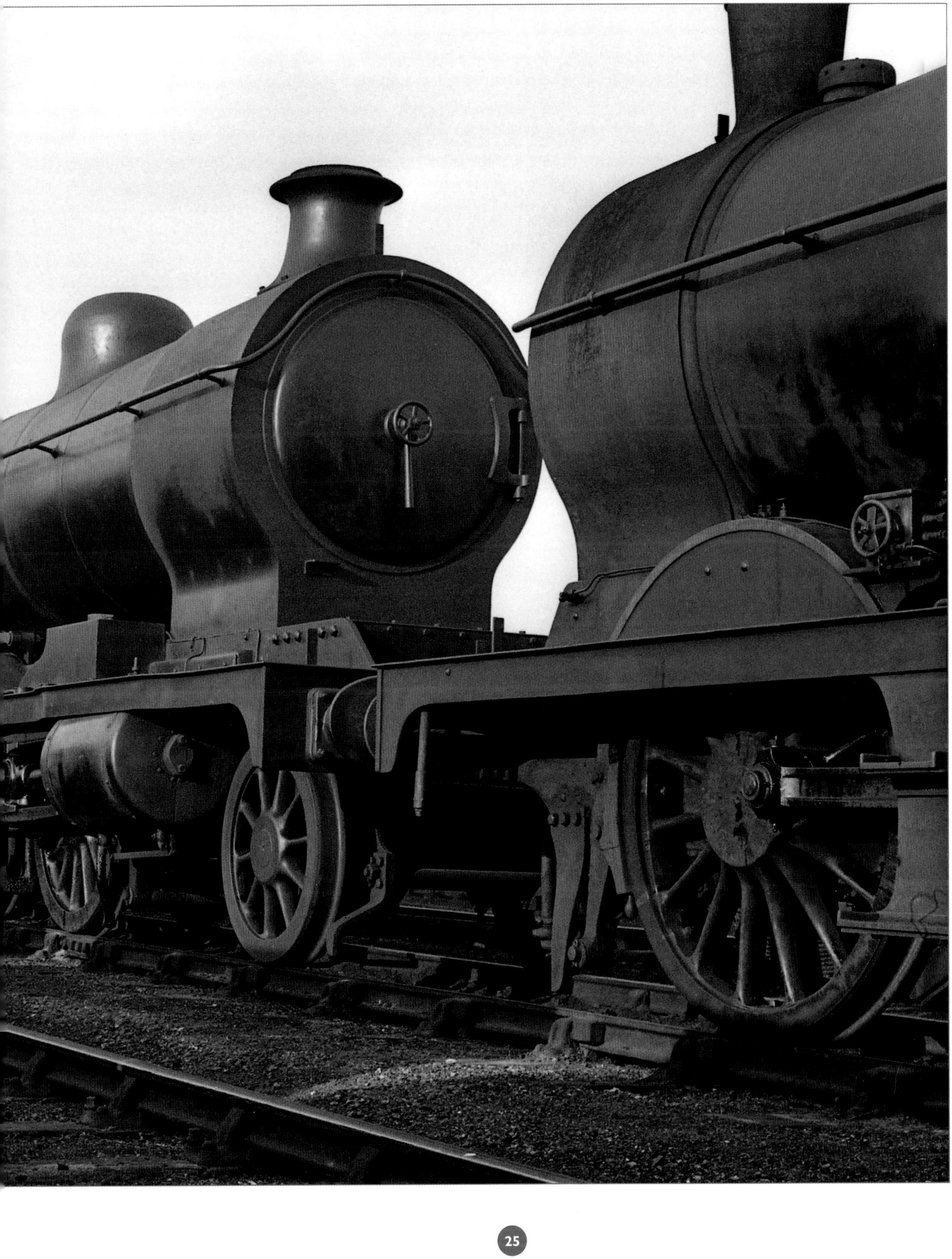

Recently buffed-up Stirling Class G1 0-4-4T No. 3766 is clearly the apple of this gent's eye as she poses in the shed yard during August 1926. One of a small class of twelve, the loco was once employed on the GNR Metropolitan suburban lines out of London until displaced by Ivatt's 4-4-2T and N1 0-6-2T types. Migrating further north, by this date No. 3766 was the surviving member of the class and clearly cherished by the shed staff. Built in December 1889 as GNR No. 766, the loco is seen retaining its GNR livery but renumbered by the L.N.E.R. – the only member of the class to be so treated. This guise shows a tall built-up chimney and a flat roofed cab which survived until February 1927 having outlived her compatriots by some years – although No. 824, withdrawn in June 1924 – survived at Doncaster as a stationary loco until 1932. *Photo: Neville Stead Collection © Transport Treasury*

Nevertheless, in railway terms at least, this was to change dramatically with the 1875 arrival of the GNR Derbyshire & Staffordshire Extension lines built to tap into the considerable mineral traffic emerging from the Derbyshire coalfields. From a population of around sixty in 1801, by 1891 Netherfield had grown to almost 2,700 souls and by 1901 to 4,646 – eclipsing that of Colwick itself and housed mainly in rows of Victorian terraces sitting astride the north-east chord of the triangle of GNR lines and south-east of the MR Nottingham to Lincoln line on which the company opened Carlton station in August 1846, renamed Carlton and Gedling in 1871 and (confusingly) from 1896 until 1973 Carlton and Netherfield for Gedling and Colwick.

Netherfield became a township and separate parish in 1885, but despite this hike in importance the yards and loco shed continued to be known as Colwick, probably as a result of the opening of the nearby GNR station of the same name in 1878, although the latter was changed in 1883 to Netherfield and Colwick!

In 1870, anticipating the opening of the Derbyshire & Staffordshire scheme, the GNR purchased around 150 acres of land for a marshalling yard, loco depot, workshops and housing for staff accommodation. The first mention of the loco depot *per se* dates from early 1875 and by the following spring a brick building with two hipped roofs covering four dead-end roads had been erected by J. Parnell & Sons to house twelve locos plus a two-road workshop. A 45ft turntable and water tower were also provided. Also opened at much the same time were adjacent sidings capable of processing 650 full wagons, 500 empties and the first rows of company staff housing. The latter comprised twelve properties each for the Loco and Traffic departments built by R. Stevenson at each side of the rear of the loco shed facing Netherfield Road and appropriately named Locomotive Terrace and Traffic Terrace.

EXPANSION

By 1881 and with the part-completion of the GNR Leen Valley line serving associated collieries producing further mineral traffic, such exponential growth was beginning to stretch the ability of Colwick shed to keep pace with

This 1923 shot captures one of Colwick's 'Baltic' Class L1s having recently acquired its L.N.E.R. livery but not its new number (3133) which was not applied until August 1925 thus making it a full member of the L.N.E.R. Class R1. No. 133 was built in December 1905 as one of a class of forty-one intended for Metropolitan duties but found to be unsuitable due to weight of around 70 tons and also (allegedly) because the class was too slow for the smartly timed services. However, No. 133 was one of the later series of the class never to have seen service on passenger trains in the London area but in 1919 was one of six sent to work empty stock workings between Kings Cross and Hornsey. By the Grouping No. 133, then based at Colwick, had been rebuilt with a larger 4ft 8in. boiler in 1909 and is seen 'on the road' near Basford (GN) in sparkling condition with a local freight. As mentioned in the text, the locos were also used from Colwick on the heavy Annesley trips but the arrival of the ex-GCR O4s spelled their end and all the class had gone by early 1934 with No. 3133 having succumbed much earlier in July 1928. *Photo: Neville Stead Collection © Transport Treasury*

demand. Patrick Stirling (the GNR Loco Superintendent since 1866) recommended an extension to the shed sufficient to house forty extra locos plus a repair shop and this work was completed by the firm of E. Wood during June 1882 for a bid price of £11,310. The eight-road 'northlight' extension (promptly dubbed 'Big Shed') was built next to the existing four-road building (now known as 'Old Shed') and at the same time a larger coal stage was added to the layout to enable replenishing of locos visiting either shed. Further sidings were added to the marshalling yard in 1884, 1888, 1889 and 1891 and in the following year Stirling was once more recommending to the GNR Board a further increase of the shed's capacity which would result in the building of a new four-road shed adjoining the western side of the 'Big Shed' and the extension of each of the existing roads in 'Old Shed', 'Big Shed' and the workshop. The new shed was naturally dubbed 'New Shed' and the whole project (now with Henry Ivatt at the loco helm since 1895) continued apace and eventually included the provision of twenty-nine extra sidings in marshalling yards, a new 175ft x 55ft double-sided 'ramp' coal stage with three coal chutes each side and two sets of internal rails capable of holding sixteen 20 ton wagons, a 55ft turntable plus extra sidings in the loco yard – all of which appears to have been more or less completed by the firm of Dennett & Ingle during 1901. The additional sidings laid in at the marshalling yard brought the total up to sixty-eight Up and sixty-seven Down enabling the processing of 6,000 wagons and was the last enlargement of its capacity and, as one of the largest freight yards in Europe, proved more than sufficient to take it through to the GNR and L.N.E.R. periods until the downturn of mineral and general goods traffic during the BR period.

HALCYON YEARS.

As the size of the shed and locos allocated increased so did staff numbers and by the end of the Great War between 900 and 1,000 of all grades were employed including around 800 footplatemen covering over 1,000 weekly duties. In Roger Griffiths' and John Hooper's authoritative account of the shed published in Volume 2 of their GNR shed series (Challenger Publications 1996) these numbers were said to have been regularly exceeded with no less than 1,379 weekly duties (or over 200 on each weekday) spread across the allocation of some 263 locos – some workload! At this time Colwick marshalling yard was probably at its busiest with around 400 staff employed there and, allegedly, on one busy day in November 1913 over 200 trains were dealt with – not just coal workings, but other minerals such as iron ore and a wealth of general

This 1962 view of the shed yard perfectly captures the essence of workaday Colwick through the years. On days such as this, which had little or no wind and the sun struggled to break through its smoky mantle, one could taste and almost touch the atmosphere while the locos themselves seemed somehow diminished within the murk. Nevertheless, such atmospheric sights, despite breaking all the 'clean air' rules of the present, are much missed by those imbued by the steam railway ethos – but not by the housewives of Netherfield who battled daily to keep their washing clean! *Photo: Canon Alec George © Transport Treasury*

L.N.E.R. Class Q2 0-8-0 No. 3416 was one of fifty-five designed by Henry Ivatt and introduced between 1901-09 for the heavy London coal traffic. Superheated No. 416 (GNR number), seen at Colwick during the 1930s, appeared in January 1903 and was one of fourteen built with 21 x 26in cylinders, piston valves, and classified as 'Q2' by the L.N.E.R. to differentiate from those of the class with slide valves and 1in. smaller cylinders. The high pitch of the boiler was typical of the type and its length had invited the 'Long Tom' nickname due to its similarity to a 155mm Creusot 'Long Tom' artillery piece used during the Boer War. Withdrawals of the class started fairly early during the late 1920s although our example lasted until December 1935 and was used mainly on local colliery trips and also on the New England – Colwick goods pick-ups via Spalding and Sleaford. *Photo: Neville Stead Collection © Transport Treasury*

During 1935 Colwick had fourteen members of the ex-GNR J13 0-6-0STs (L.N.E.R. Class J52) – a number which held fairly steadily until the mid- to late-1950s when local shunting turns had started to diminish and shunters such as the one seen here had also been whittled down by diesel replacements. Confirming this, in early 1958 a Railway Observer correspondent reported around twenty-five locos in store or out of use at Colwick. This undated shot of No. 68871 shows the loco in a typical work-weary state and doubtless not so different when she had been based at Ardsley shed since at least 1945. Built in October 1905 she had arrived at Colwick from Ardsley in January 1956 and was destined to last until withdrawn in February 1958. Originally numbered as GNR No. 1272, renumbered by the L.N.E.R. to 4272 and then 8871 in 1946, the loco was one of the last three of the class to survive at Colwick, going in February 1958 with classmate No. 68829 while the last of the trio – No. 68863 – was condemned three months later. *Photo: Neville Stead Collection © Transport Treasury*

COLWICK TRACK PLAN

- Stoke Dyke
- Footpath
- Engineers Office
- Store
- Platform
- Coals
- Engineers Yard
- Weigh Machine
- Platform
- Store
- Sand Shed
- Water Column
- Coal Stage
- Ash Pit
- Coal Stack
- Water Column
- Ash Pit
- Water Column
- Ash Pit
- Platelayers Hut
- Goods Shed
- Furnace
- Engine Pits
- Engine Pits
- Shunters Hut
- Shunters North Cabin Signal Box
- Netherfield Rly Club
- Bowling Green
- Settle Alley
- Stores
- Mess Room
- Wagon Repairing Shop
- Engine Shed
- Engine Shed
- Erecting Shop
- Traffic Terrace
- Victoria Road
- Offices
- Office Stores, etc.
- Engine Shed
- Goliath Crane
- Fuel Sheds
- Stores
- Sidings for empty wagons
- Engine Line
- Shunters Line
- Up Yard Line
- Down Yard Line
- Down Main Line
- Down Slow Line
- Main Line
- West Group Sorting Sidings

goods and perishable workings. With Gresley taking up his post of Chief Mechanical Engineer in October 1911, Colwick saw some improvements including a change from cold water to hot water boiler wash-outs with a saving of seven hours per loco, a gas producing plant from which electric power could be derived and the acquisition of a 15 ton breakdown crane as a replacement for the 8 ton hand crane which dated back to 1876. Quite why a shed of this importance had to wait so long for modern breakdown equipment puzzled Messrs Griffiths and Hooper, and indeed, it was odd.

A further workload fell to Colwick with the December 1932 closure of the nearby ex-LNWR 8-road Netherfield & Colwick loco shed and the transfer of its staff to Colwick to work coal trains to London. This shed probably dated back to the early 1880s with the completion of the GNR/LNWR Joint Nottingham–Melton Mowbray–Market Harborough line and had its own 'L' shaped terraced dwellings known as Northwestern Terrace. After the removal of the locos the shed was used as a wagon repair shop – this being moved from Colwick shed, thus freeing up space for a stores facility there.

The very last improvement under the auspices of the GNR was approval in 1922 for extending the ashpit facilities, although this job was left to the L.N.E.R. to complete along with the installation of a wheel drop at the rear of 'Old Shed' No. 1 road in 1925. In 1936 when the shed's allocation stood at around 170 locos the L.N.E.R. undertook a large modernisation project for the shed and approved the expenditure of £35,000 for the provision of a 500 ton mechanical coaling plant, a 70ft vacuum-operated turntable replacement for the existing 55ft version, a 260ft wet ash pit with ten travelling cages and a 3 ton steam grab crane, an artesian well to supply the shed and its facilities, the railwaymen's houses, a new set of water columns, and lastly, a new oil store and commodious mess room (said to be one of the best on the system) plus a new foreman's office. All these additions and improvements took some while to complete and were not finished until the early years of WW2, but more or less in time for the vast increase of war-generated work from 1941.

THE LOCO ESTABLISHMENT

It didn't take long for the growing numbers of locos to throw their smoky presence over the area and, as one resident noted, forming a yellow smog or mist mixed with engine smoke which frequently blanketed local houses. This atmosphere would persist until the official closure of the shed to steam on 12th December 1966, although in the months previous to this and with a declining number of locos hopefully the air had cleared considerably!

For the first twenty-five years or so the shed was host to Archibald Sturrock's 2-4-0 and 0-6-0 motley designs; including his 116 and 168 goods loco classes which Colwick, being predominantly responsible for freight, made good use of. It seems likely that Colwick also saw use of the Sturrock six wheel steam tender – an auxiliary engine designed to give extra power at starting and low speeds plus enabling haulage of 450 ton trains on the heavier gradients between Peterborough and London. These tenders were fitted to his 0-6-0 goods locos. With Patrick Stirling assuming control of loco matters uniformity of design was achieved with his own 2-4-0 E2 and E3 classes built between 1867 and 1895 and from 1867 his first Standard Goods 0-6-0 design which was destined to form the basis of further development of the type until the turn of the century. The latter fell into the GNR classes J7 (35 built between 1867-73) and J6 (160 built between 1873-96). Stirling was also able to rebuild selectively some of the Sturrock designs. Ivatt perpetuated Stirling's work, bringing out his own 'Standard Goods' Class J5 of which 143 were built between 1896 – 1901 which became L.N.E.R. Class J4. From 1908 Colwick also acquired from new three of Ivatt's fifteen-strong Class J21 for fast goods work to London. Later reclassified 'J1' by the L.N.E.R., in 1945 Colwick had all fifteen of the class allocated until withdrawals and transfers whittled the number down until the last on the books, then BR No. 65002, was withdrawn during August 1954.

Often overshadowed by their bigger and beefier brethren, the Stirling and Ivatt Class J13 and J14 0-6-0STs of the GNR (L.N.E.R. Classes J52 and J53), introduced from the 1890s for shunting duties, were never in great abundance at the shed until the mid-1930s when fourteen were present, eighteen in 1948 with the last being withdrawn during 1958. Synonymous with Colwick were the Ivatt Class K1 0-8-0s (L.N.E.R. Class Q1/2/3) introduced between 1901-09 to haul up to sixty loaded mineral wagons from the colliery regions to London. Dubbed 'Long-Toms' due to their long boilers, the 55-strong class proved very successful and were able to relieve existing Stirling 0-6-0s which by now were struggling to manage increasing mineral loadings. However, their demise was hastened

25th September 1949 • As one of Henry Ivatt's 'second string' 4-4-0s (see Eastern Times Issue 3), war-weary D3 4-4-0 No. 2148, seen against the soaring backdrop of the 1930s mechanical coaler, was one of many of the class to reduce in numbers after the closure of Nottingham London Road depot in c.1906. The type maintained a long association with Colwick until the early years of BR, the last not going until June 1951. No. 2148 was the penultimate member of the GNR D4 class built in December 1899 as GNR No. 1359 but rebuilt to a D3 with a higher pitched 4ft 8in boiler in November 1912. Renumbered successively 4359 and 2148 by the L.N.E.R., the loco had only been at the shed since the previous May having been at Boston for some twenty years. It never assumed its BR-allocated number (62148) and was withdrawn during December 1950. *Photo: Neville Stead Collection © Transport Treasury*

4th April 1954 • Emerging from the murk, Gresley's Class J39 0-6-0 No. 64827 was one of the 289-strong class which although first appearing at Colwick in numbers from 1936 never became a staple at the shed due to the shed's already ample stable of 2-6-0s and 0-6-0s. In October 1937 there were nine on the books and generally regarded as 'maids of all work' capable of performing not just local and medium distance goods works but also filling in on stopping passenger services and East Coast summer excursions. This example of the class had arrived from Immingham in April 1947 and stayed until moved to Doncaster in March 1959 where it was withdrawn during the following February. *Photo: Eric Sawford © Transport Treasury*

Gresley 2-cylinder 'Ragtimer' Class K2 (his first design for the GNR) No. 61726 had emerged from Doncaster during March 1913 as Class H2 No. 1636 with a 4ft 8in boiler, becoming Class K1 No. 4636 under the L.N.E.R. in 1923. She was rebuilt with a 5ft 6in diameter boiler and an improved cab to become Class K2 in July 1937 and was renumbered to 1726 in October 1946 at a time when Colwick had fifteen of the class – a number that would increase to nineteen in 1954 and illustrative of their popularity at the shed. No. 61726 shows off her energetic lines outside the repair shop on a clear day during the 1950s. She had been a Colwick resident since August 1946 having arrived from Doncaster and survived until May 1957. That they had a good turn of speed was never in dispute and confirmed by W. A. Tuplin who noted a member of the class prior to WW2 managing the 19.8 miles between Doncaster and Wakefield at 20 minutes 20 seconds whilst hauling nine carriages at around 280 tons loaded. *Photo: Neville Stead Collection © Transport Treasury*

following the mass introduction of the Robinson GCR O4 2-8-0s from 1924, but the class remained a fixture at Colwick, albeit in diminishing numbers, until the last was condemned during 1935. More or less at the same time as the introduction of the 'Long-Toms', Ivatt introduced his forty-one Class L1 0-8-2Ts, (L.N.E.R. Class R1) a tank version of the K1 and originally intended for passenger and goods workings over the Metropolitan Lines. However, the class proved less than successful on these duties due to their weight and were superseded by Ivatt's Class N1 0-6-2Ts. Most of the L1s gravitated to Colwick where they were known for some reason as 'Baltics' (in early 1914 all of the class were stationed there) and were used on local coal trains from Ilkeston, Pinxton and Shirebrook plus weightier trips to Peterborough with empty returns plus workmen's trains (or 'Paddy's Mails' as they were known locally). After the 1923 Grouping they were also used on the Colwick–Annesley–Newstead–Colwick trips which often took up to fourteen hours to complete. The first withdrawal occurred in 1927 and the class had disappeared by February 1934.

Passenger duties were never Colwick's strong suit until the closure of Nottingham London Road (Low Level) shed in c.1906 when its passenger duties to Grantham and over the GNR Derbyshire & Staffordshire line were transferred to Colwick. This resulted in numbers of Ivatt 4-4-0s of the D1 and D3 classes (L.N.E.R. Class D2 and D3) moving to Colwick with these joined from 1913 by half a dozen or so of Gresley Class H2/3 2-6-0s (L.N.E.R. Classes K1/2). Members of the latter class survived at Colwick into 1960 being regularly employed on goods services (including Burton-on-Trent beer trains) and also summer excursion traffic to East Coast resorts. One member of the class, No. 4365 (L.N.E.R. No. 2169), was present continuously at the shed from 1922 until withdrawn in 1948. The impressive ex-GCR B8 4-6-0s also appeared from 1926 until 1943 (eight of the eleven-strong class were at Colwick in 1937) and were used on fast goods duties to Hull and Manchester as well as summer excursion trains.

When the L.N.E.R. introduced the Sentinel Railcars from 1925 nine were allocated to the Southern Area and at one time or another Colwick hosted all of them for services on the Grantham line to Nottingham and also to Chesterfield, Ilkeston, Shirebrook and Sutton-in-Ashfield. Of the nine, *Commerce* had the longest tenure at Colwick from January 1929 until August 1935 returning in January 1939, and *Rival* was next, spending 4½ years at the shed. Railcar duties in the Nottingham District were withdrawn from September 1939 and *Commerce* was withdrawn in January 1940. *Rival* survived until January 1946.

As mentioned, the GCR O4 2-8-0s first appeared at Colwick in 1924 and by 1939 no less than forty-seven were allocated and the class remained a fixture at the shed until the last five on the books (Nos. 63644/74/5, 63816/73) were withdrawn following the transfer of the shed to the LMR in January 1966. Had the transfer not occurred there is probably no doubt that they would have soldiered on until the end of steam at the depot in the following December. During WW2 the ubiquitous WD 2-8-0 appeared and in 1947 over forty were on the books and had replaced the GNR O1 and O2 2-8-0s which had arrived in 1914 and 1921 respectively. The WDs remained in substantial numbers at Colwick until the shed's closure to steam in December 1966. Packing an even heavier punch, numbers of the BR 9F 2-10-0 'Spaceships' came and went during the 1960s but apart from nine withdrawn from the shed during 1965 none remained for very long and were moved on to other sheds. Gresley B17 4-6-0s and Thompson B1 4-6-0s also added their number to the Colwick lists – the former from the early 1930s until 1949/50 with these being replaced by an ever-growing number of B1s – the latter destined to be the last L.N.E.R.-built locos to leave the shed in late 1965.

The Gresley K3 2-6-0s put in their first appearance in 1937 and by 1944 there were eleven allocated at Colwick. The locos were employed on the usual bread-and-butter goods duties and also much favoured on wartime troop and leave specials. Possibly one of the class may hold the Colwick record for visiting the most far-flung location – this being No. 61894 which was noted at Port Talbot on 5th April 1961 whilst hauling a goods special destined for Milford Haven. The Thompson L1 2-6-4Ts had also appeared on the strength from 1955 and proved popular with crews with over thirty of the class allocated at different times by the early 1960s. However, all had been transferred or withdrawn by the end of 1962.

DECLINING YEARS

In April 1950 BR recoded the shed from CLK to 38A thus retaining its status as a Loco District centre with the subordinate sheds being Annesley, Leicester (GNR & GCR), Staveley and Woodford Halse with Derby Friargate remaining as a Colwick sub-shed. This arrangement lasted until February 1958 when Colwick became 40E and subordinate to Lincoln District 40A. A further (and last) change came in January 1966 when the shed became 16B and part of the LMR Nottingham District which promptly dispensed with the remaining L.N.E.R. locos (apart from B1 No. 61264 which was retained as a stationary boiler until April 1968), replacing these with a number of 8F 2-8-0s and 'Black Five' 4-6-0s. Diesel power in the shape of seven new 350 b.h.p. 0-6-0 shunters had arrived from Darlington in late 1958 followed by examples of 204 b.h.p. Drewry 0-6-0 types replacing the stalwart J52s and a few GER Class J67/9 and J94 latecomers.

Since the late 1930s the state of the shed's northlight roofing had started to cause concern but with the interruption of the war it was not until the early 1950s that BR gave it some attention. With the first attempts at replacement with aluminium sheeting proving short-lived against the deleterious effects of loco smoke, BR grasped the nettle in a more positive fashion and completely re-roofed all three 'Old' 'Big' and 'New' establishments with the tried and tested method using concrete, steel supports, asbestos sheeting with brick screening at the shed road entrances. Various other roofs such as those to the Repair Shop and former Wagon Shop were found to be sound and not replaced. At the same time as this work the original 45ft turntable located by the side of the 'Old' shed was taken up after a career spanning over 75 years.

Various 'shed bash' Sunday visits during the 1950s and '60s reflect the gradual decline in the allocation numbers

4th April 1954 • Ivatt Class J5 0-6-0 No. 65494 percolates before moving off to its next duty. Built in December 1909, it was a member of a small class of twenty (GNR J22) intended as medium powered goods locos but were soon overtaken by the introduction of the more powerful superheated and piston-valved version which became the L.N.E.R. Class J6. As descriptive of the work carried out by the J5 in later years one cannot better the racy account given on p34 of the RCTS 'Green Guide' Part 5 viz: *By 1944 the whole of the class was stationed in the Nottingham area. For many years these doughty old warriors were a familiar sight as they battled their way up the vicious gradients between the Nottinghamshire and Derbyshire coalfields and Colwick yards on a constant procession of coal trains, their peculiar, hollow-sounding exhaust disturbing the peace for miles around. Many a struggle took place to reach the summits on the outskirts of Nottingham, but they always seemed to manage it as the engine was about on its last gasp. As one of the last trio of the class, No. 65494 was the first to go in January 1955 with the other two going at the end of the year.* *Photo: Eric Sawford © Transport Treasury*

– this linked with the downturn of the marshalling yard traffic. However, this took some time to work through. Such a visit in 1954 recorded 146 locos with virtually all recorded as being Colwick residents. Noteworthy were the thirty WD 2-8-0s – all 38A residents. In June 1957 the count was slightly larger but had started to fall by July 1959 and June 1960 with visits recording 124 and 110 respectively. In May 1963 the decline had really set in with just 83 locos recorded which had further declined to just 60 in March 1965 with this number inclusive of twenty diesels.

Something of a brief and very dramatic renaissance was recorded during a February 1966 visit after the LMR 'clear-out' when no less than 133 locos were recorded – most being 8Fs and 'Black 5s' but sadly this number was inclusive of withdrawn WDs (17), B1s (11) and O4s (9). A visit three months prior to closure recorded a count of 74 – this including a reduced number of L.M.S. types, a handful of BR Class 4MTs, two visiting 9Fs and 33 diesels of various types including a visiting 'Peak' type. As a modest swansong of sorts that day, B1 No. 61264 (since 1965 masquerading in the guise of Departmental No. 29 and providing a stationary boiler) represented a solitary, tenuous link recalling the shed's fascinating past. Fortunately, the loco was lucky enough to be sold to Dai Woodham's yard at Barry – the only ex-L.N.E.R. loco to be sent there – and was rescued in 1973. Acquired by the Thompson B1 Locomotive Trust, the loco underwent a 20 year restoration plan to return to steam in 1997 and is now based at the Nottingham Heritage Railway and currently undergoing overhaul.

After the demise of steam Colwick continued as a diesel depot until final closure along with the marshalling yard during April 1970. Demolition of the shed commenced in 1971 with the coaling tower brought to its knees and rubble on 29th December 1971. Today the site of the shed is bisected with the Colwick Loop Road with offshoots to light industry and retail units although some of the terraced areas dating back to the Netherfield railway community such as Arthur, Deabill, Godfrey, Manvers, and Pearson Streets are still very much with us. Of the huge marshalling yard, this disappeared mostly under what is now the Victoria Retail Park.

Sic transit gloria mundi. (Thus passes the worldly glory.)

Sources consulted:

RCTS Locomotives of the L.N.E.R.. Various editions.
Great Northern Engine Sheds Vol. 2. Roger Griffiths & John Hooper. Challenger Publications 1996.
Rail Centres: Nottingham. Michael A. Vanns. Ian Allan 1993.
Railway Observer. Various editions through the years.
Colwick and Netherfield online archive material.

Although more tender in years than the J52s, the no less humble Riddles J94 0-6-0STs were purchased from the Ministry of Supply and added to L.N.E.R. stock during 1946-47. Seventy-five were acquired and initially spread fairly evenly around the North Eastern and Southern areas but gradually extended their modus operandi to other parts of the system. This undated shot of former Immingham residents No. 68028 and 68009 is probably post-February 1960 when the latter had arrived at Colwick but before September 1960 when No. 68028 was condemned. No. 68009 managed to survive until August 1962 by which time she was at Stratford and had been condemned during the same month.
Photo: © Transport Treasury

20th April 1963 • As mentioned in the text the sole L.N.E.R. representative at the shed after the early 1966 L.M.S. invasion was B1 4-6-0 No. 61264 seen here in happier days and in company with classmate No. 61175. Until November 1960 she had been firmly a GE section loco based at Parkeston Quay shed until moved to Colwick where she remained for the rest of her career prior to being rescued from Barry. No. 61175 had arrived from Immingham in October 1960 and remained until withdrawn from 40E in December 1963.
Photo: Neville Stead Collection © Transport Treasury

Colwick had a penchant for veteran residents seeing out their last days – often with members of certain classes becoming 'the last one' to remain on the books. Class J1 0-6-0 No. 65002 was typical of this phenomenon and was one of the fifteen-strong GNR Class J21 introduced during 1908 for fast goods workings to London. In 1945 Colwick had all fifteen (then L.N.E.R. Class J1) allocated until withdrawals and transfers whittled the number down until the last at Colwick, and penultimate survivor of the class, No. 65002, was withdrawn during August 1954. The very last, No. 65013, was withdrawn during the following November from Hitchin shed and had even appeared on Kings Cross local passenger services in the weeks leading up to withdrawal. *Photo: Neville Stead Collection © Transport Treasury*

13th October 1963 • A brooding, atmospheric shot of Class O4 2-8-0 No. 63707 taken in the shed yard. The loco was built for the GCR by Kitson & Co. during August/September 1912 numbered 1193, then becoming L.N.E.R. No. 6193 in 1924 and 3707 in February 1947. She remained essentially in an 'as-built' condition for her entire career and lasted at Colwick until July 1965 having arrived from Immingham in September 1962. The loco gained an unfortunate reputation in the early 1960s when at some date it failed to stop whilst manoeuvring within the Colwick ashpit area and propelled some wagons forcibly into the Mutual Instruction classroom and virtually demolished it. The slightly bent front end may have resulted from this altercation but then such appearances were commonly the result of heavy contacts encountered on a daily basis! *Photo: Robert Anderson © Transport Treasury*

1966 • The sad end at Colwick in early 1966 with the working L.N.E.R. residents transferred or withdrawn and replaced by numbers of ex-L.M.S. 4-6-0s and 2-8-0s. Identifiable amongst the intruders are 'Black 5' No. 45324 which had come from Banbury and stayed until August 1966 moving then to Heaton Mersey whilst '8F' No. 48609 arrived from Leicester Midland and stayed until January 1967 before moving on to Patricroft shed.

Photo: Wally Cooper © Transport Treasury

THE 'ALPINE' ROUTE LINKING HALIFAX, BRADFORD & KEIGHLEY

PART ONE

BY PHILIP HELLAWELL

Halifax to Holmfield

Halifax, Bradford, and Keighley lie on what was the frontier of the sphere of influence of the Lancashire & Yorkshire (L&YR), Midland (MR) and Great Northern (GNR) railways, so there was plenty of scope for both competition and co-operation. As regards Halifax, topographically nestling in an elevated saucer of the steep Pennine hills, it was not best placed to benefit from early railway development. The railway station, although located on the lower eastern slopes of the town, was still some 390 feet above sea level.

Consequently, much of the town's exceptional growth in the second half of the 19th Century took place on the more moderate slopes to the west and north. Textiles, confectionery and, later, engineering were the main industries in the town and the nearby Calder Valley. At first locally mined coal was the source of power to take over from the Hebble river, but much bigger and better quality supplies were essential as the industrial revolution took hold.

The families of Halifax companies run by the Crossleys (carpets) and Akroyds (worsted), along with the Fosters of Black Dyke Mills in Queensbury, realised that developing rail connections near their factories was essential for future growth. To this end, the Halifax & Ovenden Junction Railway (H&OJR) Act of 30th June 1864 gave powers to build 2¾ miles from Halifax station to the north-western suburb of Holmfield. With a capital of £90,000, the group of industrialists mentioned, plus the L&YR and GNR each subscribed £30,000. At the first meeting of the company on 30th September 1864 the Chairman, Edward Akroyd, reported that the two railway companies had agreed to work the line upon completion.

Class N1 0-6-2T No. 69474 stands at Platform 3 in Keighley station ready to leave with a train towards Queensbury.
Photo: Neville Stead Collection © Transport Treasury

Work began in 1866 but had to be suspended for a year due to complexity of construction and spiralling costs, all leading to extensions of time being granted by Parliament in 1867, 1870 and 1873. The Act of 12th August 1867 doubled the authorised capital, whilst that of 1870 led to the vesting of the line jointly into the names of just the L&YR and GNR.

To access the Northgate end of town required construction of a link line with the existing Halifax (later known as both 'Old' and 'Town') station ½ mile away and the replacement of the stone North Bridge of 1774 which trains would have to pass underneath. The engineer for the H&OJR was the highly respected John Fraser (1819-1881), one-time resident engineer of the L&YR under Sir John Hawkshaw. He was also designer of the elegant cast and wrought iron replacement North Bridge, built eleven feet higher, 60 feet in width and with twin 160 feet spans. Opened in October 1871, this iconic bridge still stands today.

By also providing passenger facilities at North Bridge, it was intended to serve parts of the town for which the existing Halifax station was inconvenient. The new North Bridge passenger station opened on 25th March 1880, albeit a fairly modest timber-built affair. However, the freight yard had a substantial four-gabled, two-storey warehouse shared by the L&YR and GNR, fourteen cranes, rail and road weighbridges, coal offices, a wagon repair shed and stables.

The GNR operated the only timetabled passenger service over the joint line to Holmfield, but the L&YR shared all the freight facilities. The first goods trains ran to North Bridge from Old station on 17th August 1874, and the first to Holmfield on 1st September 1874.

The Crossley's Carpets mill complex, a 20-acre site at Dean Clough employing 5,000 people, was located just west of North Bridge station. A most unusual rail connection to it was engineered by means of a steeply graded siding which dropped from the Down line and dived under both running lines by a sharp curve, turning 90 degrees in only forty yards, at the entrance to Woodside tunnel.

Single trucks were then winched down to the much lower level of the mill yard below and the empties hauled back up. Crossley's mill complex, with ten on-site boilers, required an extensive horse-drawn internal rail system for moving coal and other goods between the various mills and workshops, haulage later being undertaken by a Leyland steam wagon.

An idea of the challenging landscape confronting the builders is evident from the start of the tightly constrained North Bridge station which, after only a quarter-of-a-mile, leads to the 402-yard Woodside (also known as Old Lane) tunnel, which then exits onto the six-arch Lee Bridge viaduct, followed immediately by the 267-yard curved Lee Bank tunnel.

Clinging to the side of the valley, the first station was at Ovenden, opened on 2nd June 1881 which had no goods facilities and only basic passenger accommodation, since expectations of such traffic were not high. It had two timber facing platforms, the Down side having the main

A view of Ovenden station with an N1 tank approaching from Holmfield. In later years the original Up timber platform had been replaced with a concrete one with neat stone edging flags as shown here. Photo: Neville Stead Collection © Transport Treasury

THE ALPINE ROUTE

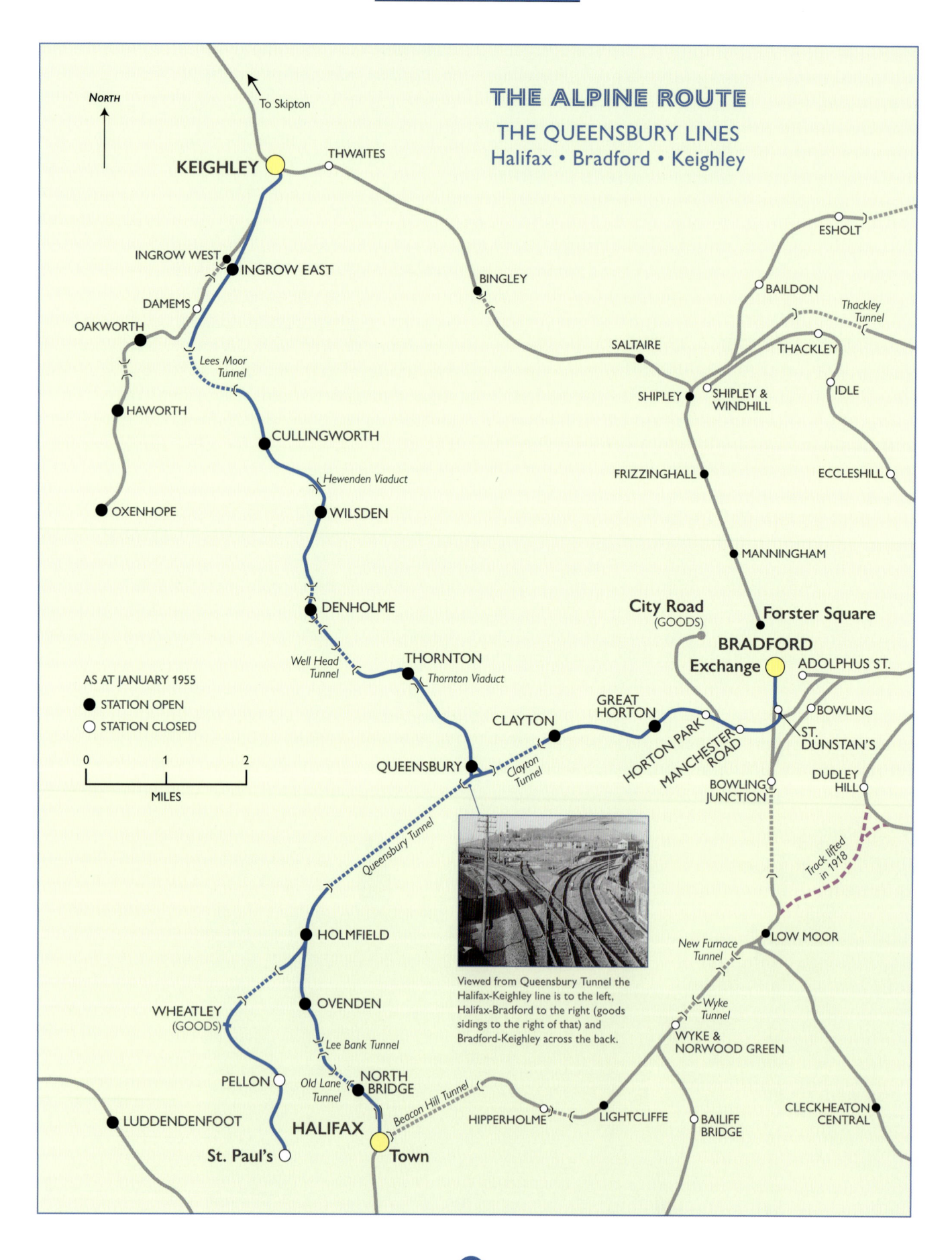

THE QUEENSBURY LINES
Halifax • Bradford • Keighley

As at January 1955
- ● Station Open
- ○ Station Closed

Viewed from Queensbury Tunnel the Halifax-Keighley line is to the left, Halifax-Bradford to the right (goods sidings to the right of that) and Bradford-Keighley across the back.

Stations and features shown on map:

Keighley, Thwaites, To Skipton, Ingrow West, Ingrow East, Damems, Oakworth, Haworth, Oxenhope, Lees Moor Tunnel, Cullingworth, Hewenden Viaduct, Wilsden, Denholme, Well Head Tunnel, Thornton, Thornton Viaduct, Queensbury, Clayton Tunnel, Clayton, Great Horton, Horton Park, Manchester Road, Bowling Junction, City Road (Goods), Forster Square, Bradford Exchange, Adolphus St., Bowling, St. Dunstan's, Dudley Hill, Track lifted in 1918, Low Moor, New Furnace Tunnel, Wyke Tunnel, Wyke & Norwood Green, Cleckheaton Central, Bailiff Bridge, Lightcliffe, Hipperholme, Beacon Hill Tunnel, North Bridge, Old Lane Tunnel, Pellon, Wheatley (Goods), Lee Bank Tunnel, Ovenden, Holmfield, Luddendenfoot, Halifax Town, St. Paul's, Bingley, Saltaire, Shipley, Shipley & Windhill, Frizzinghall, Manningham, Esholt, Baildon, Thackley Tunnel, Thackley, Idle, Eccleshill single-storey wood building incorporating ticket office, waiting room and toilets. The Up side had just a wooden waiting shelter, a signal box was added in 1900 which only lasted until the early 1930s.

An initial passenger station opened at Holmfield on 15th December 1879, which was subsequently developed with extensive goods facilities to serve the nearby engineering and woollen mills.

Holmfield to St. Paul's

Plans had been put forward in 1883 for a branch to the higher, western area of Halifax where many textile mills were being built. This would offer a service to 50,000 people and supply the needs of a large and growing manufacturing district.

Thus, the Halifax High-Level & North & South Junction Act of 1884 authorised the building of a 5¼ mile route from Holmfield through the King Cross area of Halifax to connect with a proposed extension of the Hull & Barnsley Railway (H&BR) at a central station in George Square, Halifax and thence on to Elland and Huddersfield. This envisaged grand trunk route would have allowed Anglo-Scottish expresses of the Midland Railway to travel from Sheffield through Barnsley, Huddersfield, and Halifax over the GNR to Keighley.

Ironically, the H&BR only got to Stairfoot Junction some three miles short of Barnsley, and the plan was abandoned in July 1887. A further Act of 25th September 1886 therefore limited the powers to the first 3¼ miles of the Halifax High Level Line which, due to the unforgiving Pennine topography, was constructed at the sizeable cost of £300,000. Although a short branch it involved considerable earthworks, including the 10 arch, 200 yards long, 100 feet high Wheatley viaduct over the Hebble valley, the 819 yards Wheatley tunnel, flanked by deep cuttings, numerous retaining walls of up to 70 feet high, and 16 bridges.

Nevertheless, the contractor, Charles Baker & Son of Bradford, with a workforce of over 1,000 men on the route, saw to it that, only 2½ years from commencement, the full line was formally opened on 4th September 1890. St. Paul's station was named after the local Anglican church, situated within the well-known area of Halifax called King Cross. Presumably, the unconventional naming after a parish church was to avoid confusion for passengers with the Great Northern's London terminus at Kings Cross.

The stations at both Pellon and St. Paul's were not without architectural merit. Pellon station had wooden booking offices situated on a wide overbridge at the junction of Pellon Lane and Dyson Road, connecting down to the reinforced concrete platform by a flight of steps. Timber-built waiting rooms and staff accommodation were protected by an attractive canopy of glass and ornamental ironwork.

St. Paul's was a single-storey stone-built station, with a pitched roof and raised gables. It also had a reinforced concrete island platform 376 feet long and 30 feet wide, partly covered by a glazed ridge and furrow veranda lit with gas lamps. At the southern end, a further glazed veranda with ornate ironwork covered the concourse at the head of the platform. However, only when there was an excursion was the capacity of either station fully utilised.

6th September 1953
Class N1 0-6-2T No. 69430 stands at Halifax St. Paul's station with an SLS enthusiasts' special, the last passenger train ever on the branch from Holmfield. The extent of the goods sidings, packed with wagons, give an indication of the importance of coal traffic in this yard. In the background is the tower of St. Paul's church, King Cross, after which the station was named.
Photo: Neville Stead Collection
© Transport Treasury

6th September 1953 • The SLS enthusiasts' special has reached destination its propelled by N1 0-6-2T No. 69430 from Pellon station due to the poor condition of the points at Halifax St. Paul's. The length of the platform and the ornate canopies are clearly shown which indicate how important the station was (wrongly) expected to be. *Photo: Neville Stead Collection © Transport Treasury*

Services began in 1890 with 12 return trips on weekdays and five on Sundays, all operating as shuttles. Unfortunately, it was not long before the arrival of trams in the area in 1898. This, coupled with the time-consuming circuitous route of the line and the need to change at Holmfield, caused a decline in passenger numbers.

At first, passenger trains were pulled by Stirling Class F7 0-4-2 saddle tanks, numbers 631 and 632 being long-time residents at Holmfield until the early 1900s. Nationally, the GNR started experimenting with steam railmotors in 1905, ordering six – two each from Kitsons, Kerr Stuarts and Beyer Peacock – and it is known that one such unit was tried on the High-Level line, but it was no match for the gradients. A Stirling G2 0-4-4 Well Tank was also in use around this time.

By 1910, passenger receipts were disappointingly low and on 31st December 1916, in common with many other lines the, by then, somewhat dismal service ended to release staff for war service and was never reinstated. After the war, the branch was singled and the line worked 'one engine in steam'. However, passenger trains continued to run during the annual Wakes Week and at bank holidays until 1939 plus occasional excursions to Pellon station for Rugby League matches at nearby Thrum Hall, traditional home of Halifax RLFC since 1886. The final passenger train to visit the branch was a special organised by the Stephenson Locomotive Society on 6th September 1953 hauled by ex-GNR Class N1 0-6-2T pioneer No. 69430.

Not surprisingly, goods traffic was the staple diet of the line for 70 years – during the First World War, freight included coal, khaki cloth, and army rations from the local food manufacturers. St. Paul's had a fan of four sidings to the east, and one to the north plus a coal stage, a 45-foot turntable, an inspection pit, a 2,000 gallon water tank, and a signal box on the east side of the yard.

Pellon, however, was the busier goods station, having an extensive yard with a very substantial two-storey stone-built goods warehouse and veranda, a ramped coal drop, several sidings, and two signal boxes. A petrol and oil delivery siding was added in later years. Some time after the war, traffic settled down to a basic three trains a day which continued as a regular pattern until the mid-1950s.

Holmfield, although only a village, became a busy country junction with seven sidings, a goods shed, a coal stage, a cattle dock, a 45-foot turntable, a private siding into Drake's engineering works, and one of the first oil terminals in the country. The up platform had been extended in 1890 to provide a bay on the west side for services to St. Paul's, the expansion necessitating a signal box equipped with a 70-lever Saxby & Farmer frame.

Also, in 1890, the GNR built a 120 feet x 40 feet two-road loco shed to house and service locomotives used on the branch at a cost of £4,047. Opening on 1st August it continued in use until 1933 when Sowerby Bridge depot took on responsibility for providing motive power for both the Queensbury and High-Level lines.

Bradford to Thornton

Experiencing continued frustration with the lack of progress by the Great Northern and Lancashire & Yorkshire railways, local businesses felt compelled to promote this route which, due to the considerable involvement of John Foster & Sons at Black Dyke Mills, was required to pass as near to Queensbury as possible.

The GNR, anxious to keep the Midland out and having ambition to extend the H&OJR line beyond Holmfield to meet up with a Bradford to Thornton line at Queensbury, supported this scheme and agreed to put up half the capital. Despite opposition from the L&YR, the Bradford and Thornton Railway (B&TR) Act was passed on 24th July 1871, which also authorised a goods branch to Brick Lane, 1¼ miles in length, double track throughout and known as City Road goods. The aforementioned John Fraser was also appointed as engineer for the B&TR, with his son Henry being the resident engineer, work starting in 1874. John Fraser had an impressive history, having worked under Sir John Hawkshaw, and been involved as engineer in numerous new railway lines in the West Riding of Yorkshire, in addition to several in Nottinghamshire and Leicestershire.

The B&TR was amalgamated with the GNR by a further Act of 18th July 1872 and the first sod was cut at a ceremony near Old Mill, Thornton on 21st March 1874, the line being opened in stages between 1876 and 1878. Whilst passenger services originated from Bradford Exchange, the new line proper started at St. Dunstan's to run for 5½ miles to Thornton (birthplace of all three Brontë sisters between 1816 and 1820).

Bradford Exchange

The L&YR had opened its Bradford station on 9th May 1850; described as neat and commodious it had an island platform covered by a shed 360 feet long and with a span of 63 feet. Following a connection made by the GNR from Hammerton Street junction, it was a joint station from 1st January 1867, and became known as Exchange.

Growth in passenger traffic necessitated a complete rebuild on the same site, opening in 1887/88, with ten bay platforms and two arched roofs. Constructed of wrought iron, these rested on the outside on plain stone walls and classical Corinthian columns in the middle. The roofs were 450 feet long, each arch having a span of 100 feet with a height of 80 feet, the four end screens being glazed in a fan.

Class N1 0-6-2T No. 69453 awaits 'Right away' at Bradford Exchange.
Photo: Neville Stead Collection © Transport Treasury

Class N1 0-6-2T No. 69484 at Bradford Exchange giving an idea of the size of the 1880 train shed, which had ten bay platforms and two 100 feet wide arches at a height of 80 feet from the track. *Photo: Robert Anderson © Transport Treasury*

Both companies had five platforms which were operated independently from each other, crossing over at Mill Lane junction half a mile up the 1 in 50 gradient out of the station. The two separate booking offices continued until January 1940. Surprisingly, the station never had a formal frontage, passengers entering via an opening from Hall Ings.

Exchange closed on 14th January 1973 when the City Council replaced it with a combined bus/rail station known as Interchange. Unfortunately, the rationalisation that was really needed was to combine Exchange and Forster Square stations to give a through line. So, to this day, Bradford retains its two stations in line with each other and only 300 yards apart, albeit at rather different altitudes.

In addition to Exchange, there were seven passenger stations on the route to Thornton, as follows:

St. Dunstan's

The track layout at St. Dunstan's, known originally as Bowling Junction, was unusual, with trains branching off eastwards at Mill Lane Junction and passing through a triangular junction to either run through to Leeds or, by curling back round, diving under the L&YR's route to Halifax before entering a cutting, The third side of the triangle, which allowed through running from Leeds to Queensbury, was used almost exclusively by goods trains, having no platforms as such.

5th April 1959 • Copley Hill's BR Peppercorn Class A1 4-6-2 pacific No. 60117 *Bois Roussel*, having left Bradford Exchange, heads west, passing over the GN lines at St. Dunstan's. *Photo: Robert Anderson © Transport Treasury*

Thompson B1 4-6-0 No. 61320 at St. Dunstan's, heading towards Horton Park. *Photo: Robert Anderson © Transport Treasury*

15th May 1948
Four months after nationalisation, still bearing its L.N.E.R. number and identification, Class N1 0-6-2T No. 9461, heading away from Bradford Exchange, passes signals at St. Dunstan's en route to Leeds. Note the lower quadrant signals on lattice iron post.

15th May 1948
Similarly adorned, Ardsley-based Robinson C14 4-4-2T No. 7441 passes through St. Dunstan's with a Leeds to Bradford Exchange service.

15th May 1948
Hammerton Street's Class N1 0-6-2T No. 9485 pulls away from St. Dunstan's with a Queensbury to Bradford train. Notice vintage Great Northern somersault signal.

All photos on this page by J. S. Cockshott
© Transport Treasury

A great panorama of the layout at St. Dunstan's. Left to right, the track plan is: lines to Bridge Street goods station, the ex-L&YR lines from Halifax heading for Exchange station, carriage sidings in the centre, the Queensbury lines curving round out of Exchange to pass under the lines from Halifax. *Photo: Robert Anderson © Transport Treasury*

There never was a direct passenger connection to Leeds (except for the occasional excursion trains), so passengers for Leeds or London had to change at either St. Dunstan's or Bradford Exchange.

Manchester Road

The line from St. Dunstan's climbed steeply, at up to 1 in 45 to Manchester Road station, the booking office for which straddled the railway, with a tunnel below. However, the passenger station had a very short life, being closed as a wartime economy measure in 1915.

Reopening for goods only after the war, there was an extensive yard, which was well used by the textile industry in its early days. Towards the end in 1959, it was mostly used for storage of wagons but, although it closed to freight in 1963, a single track through the site remained in use for trains serving the goods yard at Great Horton until 1972.

Horton Park

Just before this station, the freight-only City Road branch diverged, opened in 1876 on a 25-acre site with six miles of sidings. With an impressive bank of 24 double coal drops, it was located near Thornton Road, serving the fuel and freight needs of the many local textile mills. The Great Central Railway operated long distance freight trains to and from this depot, having been granted running powers.

The only passenger train to visit the branch was a Railway Correspondence & Travel Society special on 6th September 1964 comprising a BRCW 3-car Class 110 and a 2-car Metro Cammell Class 101. Horton Park did have its own coal siding, but the passenger station was never particularly busy, except when Bradford Park Avenue F.C. or Yorkshire County Cricket Club had home matches at the ground just across Park Avenue. As a consequence, it closed sooner than many other line stations, on 15th September 1952.

Great Horton

This station boasted an extensive goods yard including a parcels depot, coal drops and a large warehouse to handle trade, predominantly centred on wagon loads of baled wool linked to the nearby weaving mills. A signal box was warranted, which stood on the west-bound, Clayton, side of the island platform cantilevered out over the goods loop. Passenger-wise, the station gave a service to the neighbourhoods of Great Horton and Lidget Green until more convenient electric trolleybuses to and from Bradford lured passengers away from the railway.

This general station view of Great Horton looking towards Queensbury shows the decorative end elevation to the passenger shelter and the steps leading to the passenger footbridge.
Photo: Neville Stead Collection © Transport Treasury

Class N1 0-6-2T No. 69451 stops at Great Horton with a train for Bradford shortly before closure in 1955.
Photo: Neville Stead Collection © Transport Treasury

Clayton

The next section out to Clayton was rural by comparison, but contained a massive embankment near Pasture Lane which was 950 yards long and 62 feet deep. The small typical Great Northern island platform station at Clayton opened for goods on 9th July 1877 and for passengers on 14th October 1878. It had a substantial goods yard and warehouse which was an important freight transfer hub for separating the Bradford, Keighley and Halifax portions of goods trains.

The passenger station closed along with the rest on 23rd May 1955 and goods services ceased in 1961, but the station master's house still stands. The through line remained in use until 28th June 1965, heading north and passing under Station Road to enter another major undertaking, the 1,057-yard Clayton tunnel. This was constructed by sinking four shafts, two of which were retained for ventilation, the others being capped. Now backfilled, the entire area is covered by housing.

6th September 1953 • Class N1 0-6-2T No. 69430 tops up with water after stopping at Clayton with an SLS enthusiasts' special before heading to the High-Level line at Halifax, this was to be the last passenger train ever on the Holmfield to St. Paul's branch. The pedestrian footbridge and, by now, ramshackle signal box complete the picture. *Photo: Neville Stead Collection © Transport Treasury*

Tunnelling was dangerous and, all too often, fatal work, one tragedy taking place on 4th November 1874 when both Thomas Coates (27) and William Elliott (20) were killed at the No.1 shaft of Clayton Tunnel caused by "the neglect of the man in charge of the engine", William Francis Taylor being accused of manslaughter at the inquest in Bradford. Four men were waiting to be lowered down the shaft in a skep (basket) which was, mistakenly, drawn upwards by the engineman over the head gearing, causing them all to be ejected.

Fittingly, the two men are buried together in the graveyard of St. John's Church, Clayton with a very substantial and poignant headstone embodying this advice "Take warning at our sudden death. Make ready every day to follow us into the earth. We tell you watch and pray."

Queensbury

The most important of the intermediate towns, whose industrialists had influenced both the construction of the line and the route itself, stands 1,150 feet above sea level midway between Bradford and Halifax. Due to the shortage of flat ground, the station had to be built at the head of a Pennine valley, part filled with rubble and stone at an altitude of 750 feet.

The first platform, built in 1879, served only trains to Thornton and Bradford, but the station's importance increased greatly when the line to Halifax opened on 1st January 1890. Built at a cost of £15,000, it had six platforms, each having wooden buildings with decorative vallances, in which, apart from Ambergate in Derbyshire, it was unique in being in a triangular formation. The main station building was at the Bradford apex of the triangular layout.

To the north east the distinctive dark red brick mill chimney with three bands of enamelled white tiles has been a prominent local landmark at this point since the 1880s. The chimney belonged to the former Clayton Fireclay works and is reputed to have been built by the owner's son, Claude Whitehead, in a single day. Legend has it that the second band down, depicting a handled urn, represents the FA Cup to commemorate Bradford City's 1911 final replay victory over Newcastle United. Whether true or not, the feat, sadly, has yet to be repeated.

Connecting trains from Halifax, Bradford and Keighley often waited on each side of the triangle for passengers to change between services, the Bradford line passing through on a three-arch viaduct. Changing trains involved some variety, passengers transferring over a footbridge at the Bradford end, by boarded crossing on the Keighley side and by a vast subway on the Halifax section. Goods sidings were built behind the Bradford–Halifax platform handling coal and bales of wool inwards and finished produce outwards from Queensbury district mills.

The junction was controlled by three signal boxes, respectively known as East, South and North, although by 1944 these had been rationalised down to just the East box. Four rows of houses had been built in the township by the GNR to accommodate the navvies and their families, the streets being called Great, Northern, Railway, and Oakley (named after Sir Henry – Chief General Manager of the GNR). Although only provided as temporary homes, they actually survived until 1957.

With the station being 400 feet below the town, getting there involved a steep descent, at first involving an unmade and unlit footpath nearly a mile long. "Residents had good ground for complaint" said the Leeds Times on 11th April 1885, as traversing it on dark nights was "a trying and arduous task." Disagreement between the Queensbury Local Board and the GNR lasted many years before an access road was provided with the opening of the enlarged station in 1890. However, a visit from GNR directors in 1896 produced nothing more than a few extra lights down Station Road and a signpost and poster boards at the top.

Plans for a rope-worked inclined tramway, at a gradient of 1 in 6, connecting the town with its railhead, for which stone abutments can still be seen down Station Road, were considered in 1878 and 1887 but were abandoned. Furthermore, a rope-hauled narrow-gauge track, known as Briggs Tramway, ran from a coal mine in the valley up to the town. In 1895 it was even proposed to adapt this to facilitate passengers accessing the station, but the idea was discarded on safety grounds.

Leaving Queensbury, the line crossed the narrow High Birks valley, where a massive 900 feet long embankment over 100 feet high was constructed. Numerous attempts to build this were thwarted due to subsidence on the treacherous ground, eventually requiring 250,000 cubic yards of tipped material to consolidate.

Class N1 0-6-2T No. 69478 stops at Queensbury with a train from Halifax to Bradford.
Photo: Neville Stead Collection © Transport Treasury

Class N1 No. 69485, a long-time Hammerton Street loco, is captured here arriving at Queensbury station from Bradford with a train for Keighley. Queensbury station booking office is on the right. *Photo: © Transport Treasury*

Left: The building on Queensbury station Up platform from Keighley had clearly seen better days than when depicted here. The East signal box is visible under the footbridge.
Photo: © Transport Treasury

Thornton

Thornton station was situated at the top of a deep valley and was reached from Queensbury via a magnificent S-shaped viaduct crossing the Pinch Beck valley. 280 yards long and 104 feet high, it comprised twenty arches of 40 feet span, the piers being sunk 25 feet underground. Thankfully, this handsome viaduct survived to acquire a Grade II listing, now being preserved for the public to use as part of the Great Northern Railway Trail.

When it opened in 1878, the station was the terminus of the line from Bradford, its island platform being reached from the road by a 50 feet iron bridge, similar to Clayton and Denholme stations. Like Great Horton, Thornton, although having a busy passenger service, was more notable for its goods facilities. The depot handled coal, wood, livestock, and animal feeds and possessed a stone warehouse measuring 130 feet by 50 feet. After the Second World War, it won the best kept station award on several occasions. Following closure to passengers, the goods yard remained busy and profitable well into the 1960s, still able to attract new business.

Part Two of this article will appear in Eastern Times 5.

THE WEST RIDING

Introduced by British Railways in 1949 to replace the L.N.E.R.'s 'The West Riding Limited' which had been suspended during World War 2. The service served both Bradford Exchange and Leeds, the train splitting at Wakefield Westgate. In the early years of the service the Bradford portion was often hauled by double-headed N2 0-6-2Ts. The Down service departed Kings Cross at 7.45am, arriving at Leeds at 11.38am. The Up train departed Bradford Exchange at 7.05am and Leeds at 7.30am, the two portions joined at Wakefield and arrived in London at 11.15am. Here we see Thompson Class A1 4-6-2 No. 60131 *Osprey* in charge of the Down service entering Peterborough station on an unrecorded date. *Photo: H. Cartwright © Transport Treasury*

A FUNNY LOOKING ENGINE

BY IAN LAMB

It was commonplace five hundred years ago for murder and pillage to be carried out in the Borders 'Debatable Land' between Scotland and England that was so lawless neither country could govern it. At Kershopefoot - its epicentre - truce days sought to bring justice and peace.

But here, around fifty years ago, an execution was carried out quietly, and without ceremony, of the Minister of Transport's order to close the 'Waverley Route' between Edinburgh and Carlisle. After stalwart and determined efforts campaigning for the route's reinstatement, it has been partially reopened from the Scottish capital as far as Tweedbank. Dare we dream for further progress? Hawick beckons.

Military conflict never seems far away, and once more – around eighty years ago – this country stood alone against the might of the advancing German troops to annexe Great Britain. Initially for some this necessitated evacuation from the cities, and in my own case - not born yet - my mother, elder sister and brother found themselves on a farm in the village of Stichill, near Kelso. I do not know how long they stayed there, but the experience must have been positive as my family went on holiday there on at least two occasions not long after the Second World War had ended. Being a railwayman, my father was on a 'reserved occupation' as a wagon repairer at St. Margarets depot.

It was on one of these holidays that I recalled seeing this 'funny looking engine'. We had travelled as normal by train from the Waverley to St. Boswells before boarding the usual one coach local service to Kelso. As I excitedly looked around Kelso station, it was then that I noticed this Y1 engine. Whether it was NE 8138 or BR 68138 I cannot be certain, all I do remember is looking long and hard at this strange apparition, vowing there and then that one day I would try and model it. Like so many of my best intentions, I never got round to making that model so when *Model Rail* magazine specially commissioned my very engine, I just could not resist the temptation to purchase it.

Unkindly described in some quarters as 'a chicken hut or match box on wheels', it is an early example of Sentinel's application of their steam wagon technology of vertical boiler and high-revving geared power unit to rail use for light shunting locomotives. The 'Y1s' were single speed machines generally used for yard shunting, whilst the 'Y3s' had two-speed gearboxes enabling them to run at up to 30mph when required. A chain-drive power system to the wheels avoided the hammer blow effect of conventional locomotives.

Since the late 1990s *Model Rail* magazine has been at the forefront in commissioning relatively rare 'ready-to-run' models. This Sentinel - produced by Dapol - has sufficient weight for its size, and its performance was cannily amazing, no doubt helped with its all-wheel pick-up and drive.

Bruce McCartney - in his book *Memories of Lost Border Railways* (ISBN 9780951 785867) - drew my attention to the

1954 • Sentinel Wagon Works designed, single speed geared Sentinel locomotive Class Y1 0-4-0T introduced in 1925 weighing in at 20 tons 17cwt with 2ft. 6in. driving wheels. The parts of this class differed in detail, including size of boiler and fuel capacity. No. 68138 is pictured shunting in Kelso goods yard. *Photo: Neville Stead Collection © Transport Treasury*

25th September 1961 • One-time Western Region Class 7 Britannia Pacific No. 70018 *Flying Dutchman* at the head of the 1.28pm train from Carlisle to Edinburgh Waverley. This spot is currently occupied by a road that would need to be shared as and when the railway to Hawick is reinstated. *Photo: W. A. C. Smith © Transport Treasury*

fact that in the years before a locomotive had been allocated to Kelso, shunting was carried out using horses, nicknamed 'hairy pilots'. He also pointed out that in 1946 Kelso goods yard also had another Sentinel Y1 engine, L.N.E.R. No. 9529. In 1954 No. 68138 was transferred to Ayr, where it replaced a similar machine. For a while a Sentinel steam railcar operated on the line in L.N.E.R. days.

Apart from the main Edinburgh-Carlisle line, the railway from St Boswells to Tweedmouth via Kelso linked the 'Waverley Route' to the East Coast main line. It was a crucial diversionary path in August 1948 when floods washed away bridges and embankments between Berwick and Edinburgh. Principal expresses like *The Flying Scotsman* and *Queen of Scots* Pullman were diverted through Kelso for three months until the main line was

3rd September 1955 • Ex-NBR C16 Class 4-4-2T No. 67489 waits in the bay platform at St. Boswells with the 4.05pm train to Kelso and Berwick. *Photo: W. A. C. Smith © Transport Treasury*

open again. The line may have gone now, but British Rail must have wished that it was still there in 1979 when the Penmanshiel Tunnel collapsed causing grave disruption to trains between Edinburgh and south of the Border. Like many long-closed branches, the Kelso station complex is now a large supermarket and industrial estate.

These Sentinels were used throughout Britain from the Channel Islands to Scotland. This geographical spread of allocation made the model an attractive commercial proposition. They worked lightly laid sidings with tight curves where ordinary small shunting engines could not contemplate. Lower running costs than those of conventional steam shunting locomotives allowed the allocated sites to be served more economically.

25th September 1961 • Kelso station in its heyday. Hawick (64G) BR Standard 2-6-0 No. 78046 is about to reverse into the adjacent siding to uplift a pair of vans loaded with parcels traffic. 26 minutes were allowed for this manoeuvre.
Photo: W. A. C. Smith © Transport Treasury

The Y1/2 Sentinel 0-4-0 vertical boiler tanks were introduced between 1927 and 1933 with 15 ordered by the L.N.E.R. for shunting and light freight work. Most were withdrawn during the 1940s, but seven survived nationalisation in departmental use. The GWR also purchased one Sentinel. Others could be seen in Ireland, and further afield in Australia, Egypt and India.

It is still possible to not only see such engines in reality but to travel behind them; three survive today. Former BR 'Y3' No. 54 is on the Leeds Middleton Railway whilst sNo. 6515 *Isebrook* (former GWR No. 12) is based at the Buckinghamshire Railway Centre at Quainton Road, and the remains of the former Royal Engineers' loco *Molly* are still on Alderney. The model of No. 68138 is based on a 3D laser scan of *Isebrook*.

EASTERN TIMES • ISSUE 4

MY TRAINSPOTTING ODYSSEY – 1958
BY GEOFF COURTNEY

In a typical 1950s railway environment, No. 65511 heads an ex-Clacton freight train at Colchester. Outshopped by Stratford Works in November 1900, the Class J17, which was logged at Ilford on 18th August 1958, was withdrawn in November 1960, and the following month was noted in the Stratford scrap queue waiting to be cut up in the company of 1912-built J68 class No. 68644 and L1 class 2-6-4T Nos. 67704/26. *Photo: Dr. Ian C. Allen © The Transport Treasury*

In Eastern Times issue 3, I wrote about the first year of my trainspotting logs in 1957 at my local station of Ilford, on the former GER main line from Liverpool Street between Stratford and Romford. Not a single diesel was noted, but 1958 was a different matter, as the first signs of the new motive power that was to lead to the elimination of steam within a few years made its entry onto the stage.

Indeed, a diesel was the first locomotive to be logged as I positioned myself in Mill Road, near the Ilford flyover, for the first time in 1958, on 28th March. It was D5503 on a Down Yarmouth express, from a class that was to become an integral part of my Ilford, Stratford and Liverpool Street trainspotting days over the next few years.

The Brush Traction-built Type 2 A1A-A1A was just two months old at the time, but I don't recall regarding this interloper as the first toll of the death knell for steam. Rather I suspect my 14-year-old mind just regarded it an interesting addition to the variety of locomotives passing through Ilford, and left it to my more erudite trainspotting chums to realise the implications.

Even the appearance of D5501 hauling empty stock on the Up line didn't ring any alarm bells, for in a stay of just 1½ hours my thirst for steam was quenched by a trio of B17s, comprising No. 61660 *Hull City* on an Up Clacton, No. 61631 *Serlby Hall* on a Down parcels train, and No. 61672 *West Ham United* running light, heading, according to my notes, to Goodmayes. Throw in now preserved 'Brit' No. 70013 *Oliver Cromwell* with an Up express from Sheringham, and all was well with my world. Others noted during this stay included B1 No. 61375 on a Clacton working and classmate No. 61232 with an Up Harwich express, and No. 64662, a GER J19 class 0-6-0 built at Stratford in 1918, on a Down freight. Mention too for another pre-Grouping built engine on freight duty, No. 69498, an N2 class 0-6-2T outshopped by Doncaster in 1921 and a member of a class more associated with Kings Cross suburban trains than the GER main line.

It was to be three months before I returned to my Mill Road perch, on 24th June, and by then the diesel revolution had gained extra momentum with the introduction of powerful Type 4 English Electric 1-Co-Co-1 locomotives that had announced their arrival by muscling in on 'The Broadsman,' previously the domain of Britannia Pacifics. On this day D204, barely run-in at just a month old, came through at 3.42pm on the Down working of this flagship express, and even the most one-eyed of us east of London trainspotters began to realise that a new order was taking over.

That message had already been relayed to us 25 minutes previously when D5504 was logged on an Up Yarmouth express, and was subsequently emphasised when D202 swept past on an Up Cromer train, soon followed by D5505 heading for Liverpool Street with a Yarmouth express.

Steam still had its part to play, even if it wasn't centre stage, and during the afternoon I logged Britannia Nos. 70002 *Geoffrey Chaucer*, 70003 *John Bunyan*, 70010 *Owen Glendower*, and 70034 *Thomas Hardy*, and B17 No. 61662 *Manchester United*, all on Clacton trains, and two more B17s, No. 61612 *Houghton Hall* working a parcels train and No. 61649 *Sheffield United* on empty stock. There was also No. 61000 *Springbok* and No. 61233 on the Down and Up 'Scandinavian' respectively, so steam honour was partly satisfied.

The young fireman of No. 68538 strides forward in June 1957 to activate the Stratford depot turntable while his driver watches the photographer. A Stratford locomotive for much, and possibly all, of its existence, the Class J69, which was logged on 24th June 1958 running light through Ilford, was built at Stratford in April 1892 and withdrawn in September 1961 after a service life with the GER, L.N.E.R. and BR of just a few months short of 70 years. The smokebox numberplate from the 0-6-0T survives and is in a private railwayana collection. *Photo: Dr. Ian C. Allen © The Transport Treasury*

B1 No. 61249 *FitzHerbert Wright*, seen here in close proximity to a trackside bungalow in east Suffolk, was one of the locomotives logged by Geoff while trainspotting at Ilford on 19th August 1958.
Photo: Dr. Ian C. Allen
© *The Transport Treasury*

20th April 1957 •
No. 61535 passes Stratford close to the works where it had been built 43 years previously. The B12 was logged by Geoff at Ilford on a Down Ipswich train on 24th June the following year. *Photo: Neville Stead*
© *The Transport Treasury*

12th March 1951 •
B17 No. 61612 *Houghton Hall* makes its presence felt near Romford, seven years before Geoff noted it on a rather more humble parcels train. *Photo: J. C. Flemons*
© *The Transport Treasury*

Refreshingly, in addition to the line-up of namers and brand-new diesels logged that day, were five locomotives built by the GER at Stratford before the 1923 Grouping, two of them when Queen Victoria was still on the throne and another on a Down Ipswich express. The two oldest, both running light, were J69 class 0-6-0T No. 68538, built in April 1892 and thus 66 years of age, and No. 65446, a J15 class 0-6-0 that entered traffic in August 1899.

Another J15 also running light was 65452, built in June 1906, while J19 class 0-6-0 No. 64663 of 1918 vintage passed by on a Down freight train. The express locomotive was B12 No. 61535, which had emerged from Stratford Works 43 years earlier, in March 1915. Such commendable longevity working expresses was doubtless due to a rebuild carried out by the L.N.E.R. that included the fitting of a larger boiler.

I returned two days later, on 26th June, for an afternoon's trainspotting that included B17 Nos. 61618 *Wynyard Park* on a parcels train and 61659 *East Anglian* heading to Yarmouth, and Nos. 70003 *John Bunyan*, 70008 *Black Prince* and 70009 *Alfred the Great* on Clacton workings. The relegation of the 'Brits' from the flagship Norwich trains was abundantly clear when Type 4 D202 came triumphantly through the station on the Down 'Broadsman' en route to the cathedral city.

A quintet of B1s on Up and Down expresses, parcels and freight trains, included No. 61375 on the Down 'Scandinavian,' and one of the line's ubiquitous N7 class 0-6-2Ts, No. 69602, drifting through light on its way to I know not where (my notes said its destination was Ilford yards, but as they were way out of sight on the far side of the station, I'm thinking that was guesswork on my part).

Nearly two months later I was back, for a morning's three-hour session on 18th August, and in one period of just over an hour I logged on expresses five 'Brits', three B17s, a B1 and a K3, but not one diesel. The Pacifics were, on the Down trains, No. 70001 *Lord Hurcomb* on 'The Easterling,' and Nos. 70010 *Owen Glendower*, 70011 *Hotspur*, and 70013 *Oliver Cromwell*, all on Clacton services, and No. 70012 *John of Gaunt* on an Up ex-Sheringham and Cromer.

Meanwhile, the B17s comprised on the Down route No. 61666 *Nottingham Forest* on an express whose destination I failed to note, and on the Up line, No. 61648 *Arsenal* heading for the capital from Clacton and No. 61670 *City of London* ex-Yarmouth. The B1 was No. 61234 with an Up Clacton and the K3 No. 61810 ex-Harwich. I also noted B17 No. 61625 *Raby Castle* bound for Clacton on what I logged as the 'Holidays Express,' but I am going to have to pass on that one, for it is a titled train I can't recall. A summer extra to this popular resort, perhaps?

Other steam included J17 class 0-6-0 No. 65511, which was built at Stratford in November 1900 and therefore (by just two months) a Victorian era locomotive, while diesels comprised D5514 and D202, D204 and D205.

On the following day, 19th August, I opted for an afternoon three-hour stint, and as on the day before, steam was by far the dominant motive power on view, with diesels represented by just D204 on the Down 'Broadsman,' D5509 on a Down freight, D5500 and D5506 on Up Yarmouth trains, D5509 light, and D200 bringing in a Sheringham and Cromer express.

In contrast, the Britannia and B17 classes were still strutting their stuff, with B1 class 4-6-0s providing their usual support. Class leader No. 70000 *Britannia* came through on an Up Clacton, as did Nos. 70010 *Owen Glendower* and 70011 *Hotspur*, and No. 70013 *Oliver Cromwell* was logged on a Down Yarmouth train, meaning the two members of the class that were to be preserved a decade or so later were to pass by within a couple of hours of each other.

B17s on express duties comprised Nos. 61608 *Gunton*, 61629 *Naworth Castle*, 61649 *Sheffield United*, and 61670 *City of London*, indicating that these stalwarts weren't going down without a fight, and for good measure, Nos. 61606 *Audley End* and 61655 *Middlesbrough* showed up with parcels trains and 61672 *West Ham United* on a freight working. Of the eight B1 class engines on view, three were named: No 61001 *Eland* with a Clacton train, No. 61006 *Blackbuck* on the Down 'Scandinavian,' and No. 61249 *FitzHerbert Wright* with an Up freight.

In addition to the Brits, another Standard-bearer was Class 4MT 2-6-0 No. 76031 on freight operations, while the 19th century was represented by J69 class 0-6-0T 68549, which had emerged from the nearby Stratford Works 64 years earlier, in May 1894. This elderly lady was taking it easy running light on the Down line, but was perfectly entitled to do so as, despite her age, she still had more than three years service ahead of her.

And so my summer 1958 Ilford trainspotting came to an end, the last engine noted being B1 No. 61280 on a Down Clacton train, and I was next to visit in November. Details of these end-of-year logs, which include the introduction of a third class of new diesels, will appear in the next issue of Eastern Times.

11th April 1957 •
Looking in fine fettle, B17 4-6-0 'Footballer' No. 61662, a regular sighting by Geoff, is seen at Liverpool Street. The Manchester United nameplate and brass football are seen to good effect – what would that plate be worth at auction today? *Photo: R. C. Riley © The Transport Treasury*

26th July 1952 •
Nearly 35 years after it had been built at the nearby works as GER No. 1262, J19 class No. 64662 heads a Down relief train through Stratford. Geoff noted this 0-6-0 on a freight working at Ilford on 28th March 1958.
Photo: Peter D. T. Pescod © The Transport Treasury

10th August 1958 •
Victorian veteran 0-6-0 No. 65446 rests at Stratford MPD between duties. The 1899-built Class J15 had been logged by Geoff running light through Ilford seven weeks earlier, on 24th June.
Photo: R. C. Riley © The Transport Treasury

16th February 1957 • Britannia No. 70002 *Geoffrey Chaucer* creates a smokescreen as it waits to depart from Liverpool Street with 'The Norfolkman' to Norwich. Geoff noted this Pacific on a Clacton train on 24th June 1958, by which time newly-introduced Type 4 diesels were taking over the line's flagship expresses. On the left, adding to the the smoky atmosphere for which this London terminus was renowned, is B2 class 4-6-0 No. 61671 *Royal Sovereign*. *Photo: R. C. Riley © The Transport Treasury*

8th September 1956 • Class N7 No. 69602 heads its quad-art set of carriages away from North Woolwich station and past an estate of prefab homes. The 0-6-2T, one of a class of engines that was the mainstay of the intensive Liverpool Street suburban traffic, was logged light engine at Ilford on 26th June 1958.
Photo: A. E. Bennett
© The Transport Treasury

29th September 1951 • Class leader No. 70000 *Britannia* has a tender full of coal and is ready for the 'right away' from Liverpool Street with 'The Norfolkman' to Norwich on the day when the Pacific was to celebrate its first birthday. This Standard engine was a regular sight storming through Ilford, and one such logging by Geoff was on 19th August 1958 when it came through on an Up Clacton express.
Photo: Roy Edgar Vincent
© The Transport Treasury

21st April 1951 • Pristine Britannia No. 70003 *John Bunyan* sweeps past Geoff's favourite trainspotting location beside Ilford flyover with 'The Norfolkman' when the Pacific was just one month old.
Photo: Roy Edgar Vincent
© The Transport Treasury

No. 2395
Britain's Mightiest Locomotive
By David Cullen

L.N.E.R. Class U1 No. 2395 at Doncaster on 2nd April 1939.
Photo: Neville Stead Collection © The Transport Treasury

Picture the scene. A remote stretch of rail track in Kenya around 1960. The sun is beating down. A freight train approaches, at its head a steam locomotive, its outline blurred by the heat-haze. Wait, are there two locomotives?

No, there's only one. No... there appears to be something in between. As it nears, it pierces the haze and the mystery is revealed. It is an articulated Garratt, a red, East African Railways Class 59, with an awesome 4-8-2+2-8-4 wheel arrangement. This engine is No. 5928, named *Mount Kilimanjaro*. As it storms past, smoke billows and its safety valves roar from its superb boiler producing steam in abundance. A seemingly endless string of wagons and oil-tanks following in its wake, No. 5928 disappears back into the shimmering haze.

On the African continent, locomotives such as this once revelled in handling the heaviest loads. The type was named after its British creator Herbert W. Garratt, employed as an inspecting engineer by New South Wales Government Railways. Inspired by the carriages of rail-mounted military guns, Garratt patented his concept in 1907, subsequently approaching the engineering firm Beyer, Peacock & Company in Manchester to discuss production. This initiated a true rail icon, with the E.A.R. Class 59 ultimately holding the honour as the world's largest and most powerful working steam locomotive. Garratts were built in a vast array of track-gauges, sizes, weights, power ranges and wheel notations and operated by railways worldwide, including here in Britain.

To the Garratt design. While debatable, neither one locomotive nor two, the concept offers the advantages of both. Resembling two engines coupled, it is effectively just that with corresponding capability, while having a single boiler/firebox and only requiring one crew, is considerably less expensive to build and operate. The boiler/firebox and cab are set on a cradle-frame forming the middle section of three. This is supported by two 'engine units' incorporating the cylinders and running gear, one placed ahead of the smokebox and facing forwards, the second behind the cab and opposite facing. This makes for equally good running in either direction, eliminating the need for 'turning', which would be impractical on a standard turntable with a locomotive this length.

Vertical pivoted joints connect the engine units with the boiler section, enabling them to swing side-to-side in relation to it. This facilitates negotiation of curves and points, impossible were the engine of rigid construction. An added advantage is the stability from the main section moving laterally inward on bends, which counters the outward acting centrifugal forces. Compared with a regular locomotive, the boiler is short and of large diameter, producing copious quantities of hot, dry, highly efficient steam. In addition, the risk of the water level uncovering the firebox crown on steep gradients is almost eliminated. Of necessity, flexibility is provided in the live and exhaust steam pipes serving the engine units.

The multiple pairs of coupled wheels provide superb traction, while the overall arrangement spreads the locomotive's considerable weight to produce a low axle load. Combined, these factors give excellent hauling power and compatibility with less sturdily laid tracks. The absence of wheels and associated gear beneath the main section has a number of advantages. There is adequate space for the large boiler with its appropriately sized firebox and grate, an unrestricted air supply to the firebed ensures efficient combustion, a high-capacity ashpan permits long journeys between servicing, and the risk of falling ashes damaging running-gear is eliminated.

Like a conventional tender, the unit behind the cab holds solid fuel or fuel-oil and water. However, the water is an auxiliary load, vital in hot climates where the resource is scarce and pick-up points few and far between. The main supply is carried in a tank forming the main section of the fore engine unit. The drawback to all this is consumption of water and fuel progressively reduces the weight carried over both sets of driving wheels. Train tonnage must therefore be carefully calculated to avoid the load eventually outstripping available traction.

Many top of the range Garratts incorporate equipment such as feed-water heaters, duplicate controls in the cab's rear for running bunker first, blowers for cleaning the boiler tubes while on the move, self-adjusting pivot joints, and in hot climates, cab ventilation fans. Most Garratt cabs are well enclosed, and while providing better than average shelter for crews, they can also be prone to excessive temperatures. Locomotives of this magnitude were not generally necessary in Britain, although thirty-three 2-6-0+0-6-2 Garratts were operated by the London,

British Railways Class U1 69999 at Crowden, Cheshire (now Derbyshire's most northerly settlement) on 11th October 1953.
Photo: Neville Stead Collection © The Transport Treasury

Midland & Scottish Railway for especially demanding duties. As indeed was one remarkable machine built by the London & North Eastern. Originally conceived in 1910 by the Great Central Railway, this was to be a four-cylinder, double 0-8-0 based on the Class 8A of 1902. Subsequently redesigned as a double G.C.R. 8K (used by the military during the First World War and reclassified L.N.E.R. O4 following the railway regrouping of 1923), it took the wheel arrangement 2-8-0+0-8-2. Two were planned and funded, the order being placed with Beyer, Peacock & Company on 8th April 1924. This was amended on 31st July to a single engine. Chief Mechanical Engineer Mr. Herbert Nigel Gresley, eventually of course Sir Nigel in recognition of his achievements, then added his own input to create a double version of his three-cylinder Class O2. Construction began the following year on 1st June. Delivered to the L.N.E.R. on the 21st, something of a record considering the mammoth task, it was subsequently dispatched to the railway's Doncaster works for testing and minor modifications. The final cost of this giant was £14,895. This was the first main line Garratt and the only locomotive with this wheel notation built for use in Britain. Further, it still holds the record as the largest, heaviest and most powerful steam locomotive ever to run on British metals.

The locomotive's first official duty began on 1st July while still clad in dull grey workshop paint, as exhibit number 42 in the centenary celebrations of the Stockton & Darlington Railway. Following the festivities and two weeks on display at Darlington, it returned to Doncaster on the 18th to receive a black topcoat. Several weeks of operational trials followed. These having proved satisfactory, the locomotive began revenue service in the August. Given the unique classification U1 and running number 2395, it was initially allocated to the L.N.E.R.'s Barnsley depot, transferring to Mexborough shed near Doncaster two months later.

To some technical specifications. The boiler was a massive 7 feet in diameter by just under 13 feet 7 inches in length, containing an initial 3,581 sq.ft. of heating surface. The firebox which provided 223½ sq.ft. of this, was given a 56½ sq.ft. grate – more on this to follow. A superheater was incorporated, having 650 sq.ft. of elements. The system produced a maximum steam pressure of 180 lbs/sq.in. The locomotive's overall length was 87 feet 3 inches and total wheelbase 79 feet 1 inch. In full working order she weighed 178 tons. As a freight engine with hauling power taking precedence over speed, the sixteen driving wheels were a modest 4 feet 8 inches in diameter. The four bogies were 2 feet 8 inches. Considering the engine's size, its axle load was 18 tons 6 cwt, 14 cwt less than the L.N.E.R. N2, a relatively modest 0-6-2T passenger tank.

The U1's cylinders, six in number, were 18½ inches in diameter by 26 inches stroke, set three laterally at the outermost end of each engine unit. Fitted with Gresley's famous 'conjugation' equipment, each inside cylinder was worked by the Walschaerts valve gear of its outside units. No. 2395 boasted a Tractive Effort of 72,940 lbs, more than double that of many locomotives of the time. Although an artificially created value, as the indication of pulling power this factor is critical to rail engineers. For this locomotive tractive effort is calculated with the formula:

$$3x \frac{D^2 \times S \times P}{W}$$

in which D = cylinder diameter, S = piston stroke and W = driving wheel diameter, all expressed in inches. P = the average working steam pressure in pounds per square inch, taken as 85% of maximum. The main equation produces the figure for a simple-expansion locomotive with two cylinders, the '3x' prefix obviously doing so for this, having six. The tractive effort quoted has been rounded up to the nearest ten pounds. Coal and water capacities were 7 tons and 5,000 gallons respectively.

The U1 earned itself the epithet 'The Wath Banker', from its allocated duty providing banking (rear-end assistance), to trains assembled in the marshalling yards at Wath, north of Rotherham. Supplying Manchester heavy industry with coal from the South Yorkshire collieries, these trains would comprise some sixty loaded wagons totalling around a thousand tons. These would leave Wath between two locomotives, usually an O4 Class 2-8-0 at the head, with a second O4 or an L1 Class 2-6-4T banking. Although these two coped adequately here, this could not last.

The section of route between Wentworth Junction and West Silkstone Junction included a gruelling 1 in 40 ascending gradient stretching some three-and-a-half miles. Known as Worsborough Bank or Incline, this was far too severe for such a load to be handled by just two locomotives. Stationed in the sidings at Wentworth, No. 2395 would wait for the train to pass then move up and join the rear to provide the additional muscle. It would make up to eighteen such trips each working day. Prior to its introduction, two additional engines and crews had been required. The stint would take around a quarter of an hour, after which 2395 would detach at West Silkstone and return to Wentworth sidings until next required. The train would continue with the two original locomotives to join the main Sheffield–Manchester line at Penistone. Its work then completed, the banking engine would detach at Dunford Bridge. Despite being short, this run would have been horrendous for 2395's crew. The tunnels on the line were notorious for smoke pollution. Always on the back of the train, the men would be subjected to the pungent exhausts of three locomotives labouring extremely hard, with its massive firebox and grate, their own by far the worst. Breathing apparatus for providing cleaner air from track-level was tried, but as locomotive equipment rather than personal issue, and crews objected to sharing on health grounds. This was subsequently withdrawn and the men reverted to the traditional practice of covering their faces with water-soaked cloths in the tunnels.

On 31st March 1930 the engine was given a break from this duty when allocated the well-suited task of hauling a twenty-carriage industrial exhibition train from Hexthorpe near Doncaster to Sheffield. During its final stage, she was honoured by the presence of Sheffield's Lord Mayor on the footplate. This was to be her last outing for a considerable while however. Later that day she ran to Doncaster works for replacement of her prematurely worn-out firebox. Work did not in fact begin until October.

The locomotive suffered a number of such failings, put down to the area's water lacking the minerals necessary to form protective lime-scale on internal surfaces. Resulting corrosion had indeed made boiler retubing necessary after just a year of operation. Then in 1927, cracks attributed to the same cause were discovered in the firebox, seeing the locomotive once again in need of attention. As stated, the boiler initially had 3,581 sq.ft. of heating surface.

While the firebox was under repair, twenty of the boiler's 259 tubes were removed to strengthen the area around the flange. Its overall heating surface subsequently reduced to 3,377½ sq.ft. No. 2395 returned to work on 8th October. Corrosion continued to be a problem until the following year, when finally solved by the regular addition of a chemical solution to the boiler water.

Another recurring problem was the steam-powered reverser shifting from its setting under exertion. This caused numerous incidents, including stalling altogether during B.R. days while banking in the Midlands for another Garratt, ex-L.M.S. 2-6-0+0-6-2 No. 47972. A veteran

No. 69999 at Gorton on 18th June 1955.
Photo: Peter Gray © The Transport Treasury

An undated shot of No. 69999 at Bromsgrove beginning its ascent of the Lickey Incline.
Photo: Neville Stead Collection © The Transport Treasury

0-10-0 banking engine was sent to the rescue, culminating in the train restarting with three colossal locomotives. That would have been a truly awe-inspiring sight.

As we have seen, the locomotive's main duty was the assistance of heavy trains over a short section of route. So why did it not handle the entire run from Wath to Manchester as might be expected? The answer lies in a limitation rather than a failing, paradoxically due to the very power output for which it had been created. This came from its six cylinders. Their heavy steam requirement was generated in a larger than average boiler, heated by the firebox with its 56½ sq.ft. grate. Quite simply, this consumed coal at a rate far in excess of the amount one fireman could continuously shovel. Working flat-out during the climb over Worsborough represented the limit of physical capability. 50 sq.ft. of grate is in general considered the upper limit for manual firing. Beyond this, mechanical stoking or oil burning is generally indicated. Concerning the delay fitting the new firebox in 1930, it has been suggested this was due to looking into the fitting of a labour-saving rotary coal bunker. This could have alleviated the stoking problem but was not in the end carried out. Oil-firing was eventually tried.

In 1946 The Wath Banker was given the new running number 9999. 1948 then saw the nationalisation of the four privately owned railway companies: the L.N.E.R., L.M.S., Great Western and the Southern. It was subsequently allocated a five figure British Railways number, 69999 in the November. Despite these changes, it retained its original identity in spirit, proudly bearing small 2395 cab side plates to the end.

In 1949 an electrification programme was being planned on the Manchester–Sheffield line, which would ultimately render the Worsborough duty obsolete. At the same time the U1 boiler was nearing the end of its working life, the expense of replacement only deemed justifiable if work could be found elsewhere. Fortunately it was. The locomotive was subsequently transferred to Bromsgrove in the then B.R. London Midland Region, intended to replace an ageing 0-10-0 banker on the 2 mile, 1 in 37¾ Lickey Incline. Built by the old Midland Railway in 1919 and known as Big Bertha/Big Emma, this giant originally bore the number 2290, then 22290 under the L.M.S. By this time she was B.R. No. 58100, and in an ironic twist, the very 0-10-0 sent to rescue her stalled 'replacement.' Further, she continued working until the year after the U1 ceased.

In November 1950 No. 69999 returned to Mexborough, remaining idle until the following February when she was once again allocated the Worsborough banking duty. The issue of her massive grate was finally addressed in 1952 with conversion to oil-burning at Manchester's Gorton works.

This could have vastly improved her operation, but regrettably it was not successful. Following a lacklustre test steaming, she was again temporarily taken out of service. During this time, some cosmetic work was carried out, and on returning to duty in September 1953, 69999 bore this number on her cab sides and B.R. 'Lion and Wheel' logos on both engine units.

Following a period of sporadic working, the locomotive returned for a second spell at Lickey in mid-1955. This was short-lived. The L.N.E.R. machine evidently fell foul of the narrower Midland loading-gauge resulting in some platform damage, although this was almost incidental. It simply wasn't liked by Bromsgrove crews who complained about the great difficulty judging its length when coupling, especially in the dark. Despite the fitting of an electric spot lamp to assist, its unpopularity remained as strong as ever, though it is not impossible that inter-company and inter-regional rivalries might have had some bearing. This situation, plus the failure of the oil-firing experiment, were given as the reasons when the U1 was taken out of service on 9th October and laid up at Gorton, this time permanently. The locomotive's fate had been irrevocably sealed, as when withdrawn officially just before Christmas, it was already awaiting disposal at Doncaster. The cutters' torches carried out their grim task in the following February and March, after which nothing remained, not even the 2395 cab side plates, defiantly displayed after two re-numberings.

Despite imperfections, as Britain's largest, heaviest and most powerful steam locomotive, the L.N.E.R. Garratt was a magnificent piece of Britain's railway history. It did the work of at least two regular engines, at considerably less cost in construction, crewing and fuel consumption. Moving immense loads over two of the toughest sections of route in the country, it covered 425,213 miles in 30 years. In addition, its appearance rivalled that of the legendary Union Pacific Big Boys. Well it came close.

B.R. had their reasons for deciding its fate, but how great it would have been if instead, the engine had been earmarked for preservation, so steam enthusiasts like me might have stood in awe before it at York's National Railway Museum.

No. 69999 at Dewsnap, near Manchester on 12th April 1954.
Photo: Neville Stead Collection © The Transport Treasury

EASTERN REGION TOTEM SIGNS – THE SCARCEST OF THEM ALL

BY DAVE BRENNAND

The birth of British Railways in 1948 saw a radical and long overdue change in the Nation's run down railway system. The unification of the 'Big Four' companies created the need for a new logo to entice the travelling public back onto the railway after the post-war period of decline.

The now familiar BR totem emblem was a simple eye-catching design and quickly appeared on all publicity, station signs and railway-owned road vehicles. The designer was Mr. A. J. White from The Publicity & Public Relations Department. The use of the totem as station signage quickly evolved and spread across the UK network in the 1950s, but not every station was equipped with them. At its zenith the UK system consisted of 6,747 stations, but just 2,720 were equipped with totem signs. The regional colours adopted were London Midland Maroon, Western Region Chocolate & Cream, Southern Region Green, Eastern Region Dark Blue, Scottish Region Light Blue and North Eastern Region Orange. Much has been written about these signs over the last 30 years and I have had an interest in them for the last 45 years. This short article looks at the most difficult line in the UK to collect BR totem signs from; the Shenfield to Southend-on-Sea Victoria line in Essex.

I've come to this conclusion as just one original enamel totem sign from the whole line, namely Wickford, has ever come to light since they were removed circa 1960. Shenfield station was not fitted with totem signs, but they could be seen at Billericay, Wickford, Rayleigh, Hockley, Rochford, Prittlewell and Southend-on-Sea Victoria. The reason for their scarcity and early demise was the installation of fluorescent lighting along the route. The line was an early candidate for electrification and the Liverpool Street to Shenfield line was electrified by 1949. It was a further seven years before the branch to Southend Victoria was fitted with overhead power. Services until 1956 were entirely steam hauled and the changeover to electric traction so quickly was an engineering triumph at that time. A similar traction changeover occurred on the North East London suburban lines to Chingford and Enfield Town, where N7 0-6-2 tank engines dominated the scene for decades prior to the electrification which was completed in 1960. Totem signs from these lines are also very scarce due to the installation of fluorescent lighting during the early 1960s with the station name on the perspex light cover. Only one or two – Seven Sisters, White Hart Lane, Lower Edmonton and Bush Hill Park – totems are known

The view looking towards Shenfield in 1959 with both running-in board and totem shown to good effect.

to survive at the time of writing. Enfield Town did not have them fitted. On the Chingford line just one example each from St. James Street, Wood Street and Highams Park are known to survive and Chingford did not have totems. These signs are hugely collectable today in stark contrast to the way they were treated as worthless junk by those that removed them!

The great mystery is what happened to so many dark blue enamel signs that were removed over a period of several years? Taking the number of stations fitted with totems on this line into account, particularly the fact that four platforms at Southend-on-Sea Victoria had them, the total number fitted to this GE branch line could be between 80-100. Not all stations on the Southend Victoria branch were resigned at the same time. An interesting fact emerges when looking at the three stations in Southend that had totem signs fitted, including the two on the London, Tilbury & Southend line: Southend-on-Sea Central and East. Just one survivor (a Southend-on-Sea Central) has ever appeared on the open market. This is astonishing when considering that the total number fitted to just these three large stations could have been approximately 50-60 signs. Some were dealt with in the late 1950s and almost new totem signs were dumped unceremoniously, probably still attached to the cast iron lamp posts they were attached to. Photographic evidence shows that some stations retained their totem signs in 1961, the final year of the cull. There were very few collectors of such memorabilia in the early 1960s. Totem signs then would have been seen as relatively modern and not worth saving, certainly by the BR managers, who let valuable items (in today's market) go for scrap. There was a real demand for redundant steam locomotive nameplates during the same period and the majority of the B17 Sandringham/Footballer class plates are known to survive. These were frequently seen at the head of Liverpool Street to Southend Victoria services prior to electrification.

The surviving Wickford totem only came to light in 2018. It had been saved in the early 1960s by an elderly gentleman in Cambridgeshire. It now resides very close to its original location at Mangapps Railway Museum in Burnham-on-Crouch, Essex, which houses an astonishing array of memorabilia, locomotives and rolling stock. The museum also has a Burnham-on-Crouch totem sign on display – the only station fitted with them on the entire Southminster branch line. Hopefully, somewhere out there, more totems from this popular seaside route in a mostly rural part of Essex have escaped the scrapman's clutches. Please let us know if you find any!

BRITISH RAILWAYS STATION TOTEMS – THE COMPLETE GUIDE by D. Brennand & R. Furness was published in 2022.
ISBN No. 978-1-80035-141-7.
Priced £30 and available from Crecy Publishing.

A delightful nostalgic mix in 1956.

Southend-on-Sea East

EASTERN TIMES • ISSUE 4

DELTIC DAYS AT KINGS CROSS
BY ROGER GEACH

2nd March 1968 • D9015 *Tulyar* at Kings Cross, the steam heat boiler obviously working well! The view looks towards Gas Works tunnels, the stabling point is out of view on the left. Note the eight mile an hour cut out speed restriction signs.

EASTERN TIMES • ISSUE 4

Living down in the West Country, I was used to seeing Western hydraulics appear from around the corner, any thoughts of the Eastern Region were just that, and the opportunity to visit such far away places as London Kings Cross were very limited to me as a youngster.

The opportunity arose though, while on a Western Region Rover during the summer of 1971, to take the Circle Line from Paddington to London Kings Cross.

I will never forget the noise or the sight on emerging from the Underground and seeing a Deltic stood on the buffer stops at Kings Cross.

Like all enthusiasts at that time we made our way to the country end of the station opposite the large signal box and facing the locomotive stabling siding. There was plenty to observe and see what was going on. I believe this was platform 10 at the time but later renumbered. I may be wrong as it's over 50 years ago now.

There were plenty of light locomotive moves from the depot sidings and servicing point back and forth from the station. A Class 31 arrived on one side of the station, disappeared down a black hole only to reappear sometime later on. These were what I called block enders to the other side of the station, all fascinating stuff. They were Mark 1 non-gangway passenger vehicles that served Moorgate and Farringdon via the City Widened Lines. The local DMUs were also seen but totally ignored at the time as there were far too many loco moves to take too much notice of them. York Road platform was situated after exiting Gasworks Tunnel with passenger access off York Road alongside Kings Cross Station, this was roughly situated close to the new Kings Cross Power signal box.

There were a great number of Class 31s to be seen, often over and over again during the day. Finsbury Park had a large allocation at the time as they were used on local passenger services, empty coaching stock duties, parcel trains and freight. You would see class 40s, class 46s, a great deal of class 47s and Deltics.

Eastern Region Class 47s were not seen too often down in the West Country so it was interesting to see the Class 47s in the number series 1100 to 1111, these were York and Leeds Holbeck based engines. Scottish Region allocated 47s also appeared, mostly off overnight services, and a load of Eastern Region 47s from Finsbury Park and all over the region.

While you could see the odd London Midland Region based class 40s they were mainly the Eastern Region boiler fitted engines that appeared from York and Gateshead. The Scottish allocated class 40s also appeared off the overnight services. Class 46s were generally Gateshead based engines and were common on mail trains and overnight services.

Class 37 locos were at this time very rare visitors, unlike in the 1960s when they were far more common working from Sheffield. Likewise class 25s were very rare although Peterborough drivers had traction knowledge so they did occasionally arrive with a Class 31, particularly on Peterborough commuter services.

Electrification of the lines from Moorgate and Kings Cross as far as Hitchin on the main line and Royston on the Cambridge line was authorised in 1971. The route was also re-signalled with a new power box situated at Kings Cross which opened in stages from September 1971. Moorgate to Drayton Park was electrified by 750v DC third rail, the Kings Cross to Royston route with 25 kV AC overhead wires, including the Hertford North loop. A new Electric Multiple Unit depot was built at Hornsey where class 312 and 313 units were based. Electrification commenced with the inner suburban lines from Welwyn Garden City and Hertford North during 1976 with the route to Royston from 1978; known as the Great Northern Electrics, they transformed the railway.

Above right: **26th November 1974** • A classic view from the top end of Kings Cross looking at the stabling point. The service shed is behind the rear cab of 40086. This loco was at the time based at York depot and very much an Eastern Region locomotive. It was boiler fitted and thus quite common on passenger services, both daytime and overnight trains. On the left is Gateshead allocated Class 46, No. 46054 with the depot fuel tanks stabled behind.

The Kings Cross National Carriers depot towers above the stabling point, very much part of the scene then. How this view has changed in very recent times (2023) with the building of an enormous office complex. *Photo: © Roger Geach*

Below right: **14th August 1977** • Deltic No. 55008 *The Green Howards* stands at Kings Cross with the 09.05 service to Harrogate. It appears that the train crew are in discussion with the Station Supervisor. Electrification has taken place but the old oil powered style tail lamps still reign supreme, a couple can be seen foreground on the platform.
Photo: G. H. Taylor © Transport Treasury

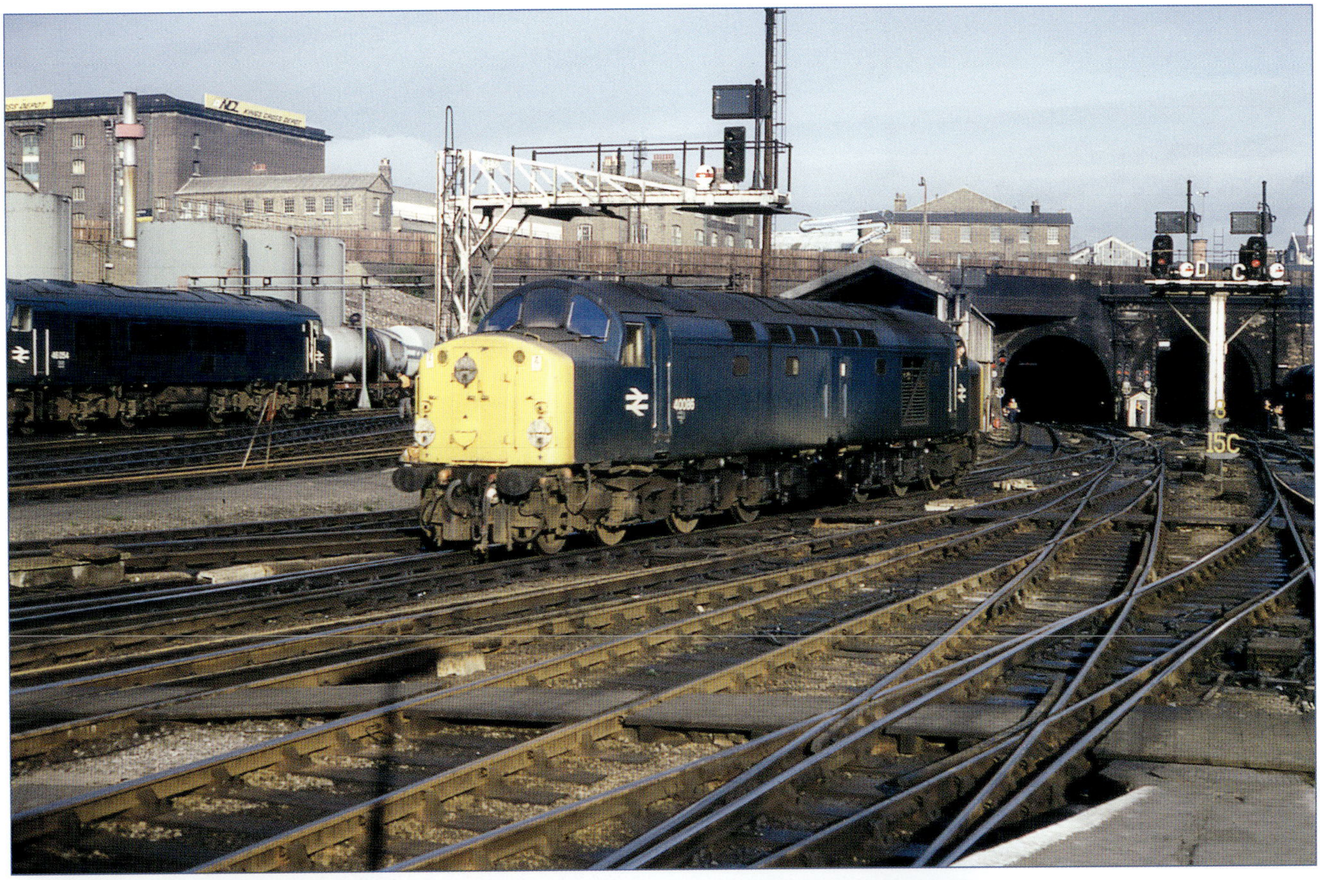
15C

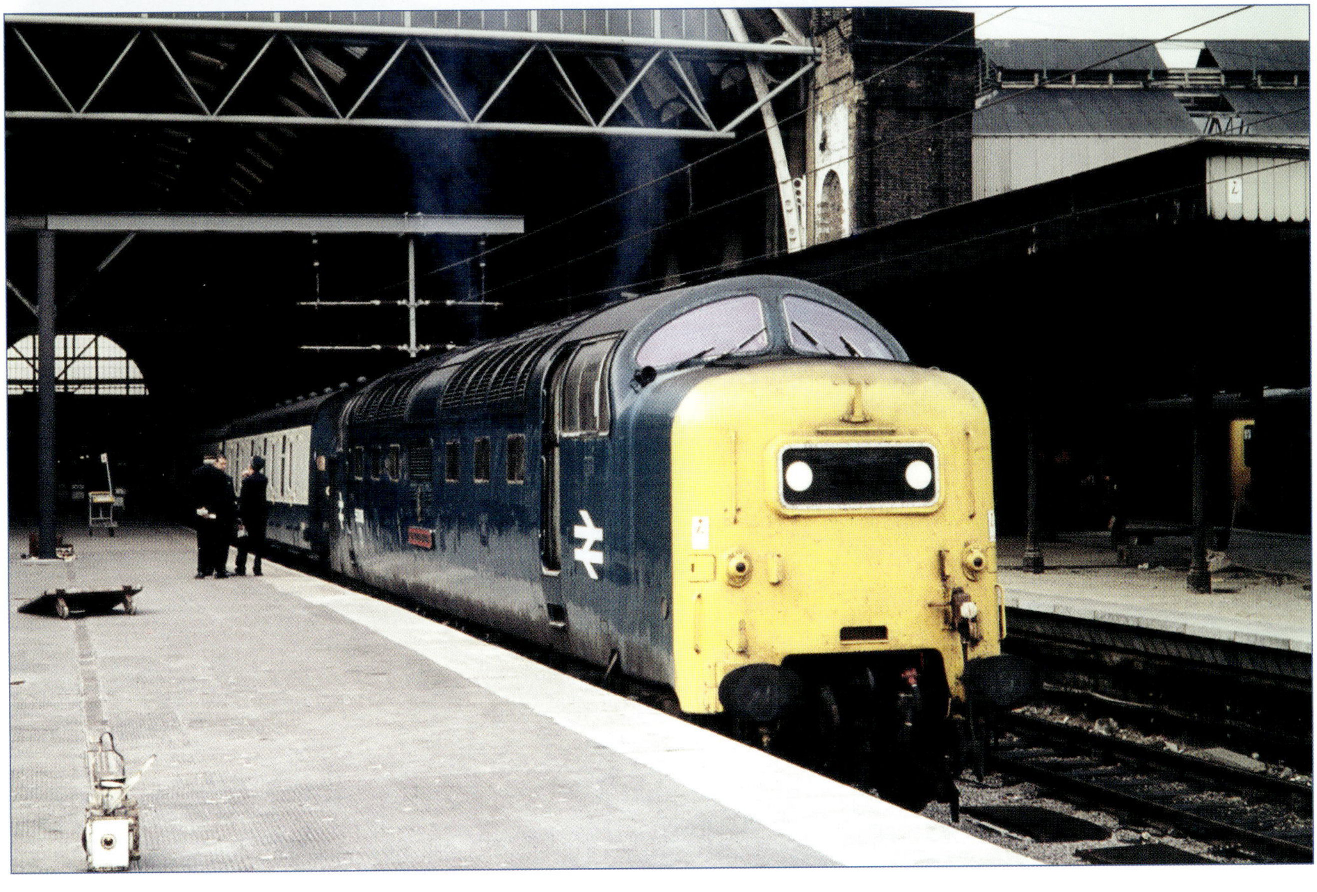

June 1971 • No. 9019 *Royal Highland Fusilier* is stood at the London end of the stabling sidings. There is a lot of lost character in this railway scene, note the 2 m.p.h. speed restriction and the red sign, along with the yellow ground position signal. Depending where one came from they were known as 'Dods' or 'Dolly' by train crew. No. 9019 had worked in on 1A04, the 07.15 from Bradford.
Photo: Roger Geach Collection

11th March 1978 • No. 55006 *The Fife & Forfar Yeomanry* ready to depart with the 13.00 Kings Cross to Edinburgh service. The new High Speed Trains were gradually being introduced into service at this point therefore replacing the Deltics. This was particularly so on the Anglo-Scottish workings. *Photo: © Roger Geach*

29th March 1974 • The signal box at Kings Cross was a big feature in its day, it towered over the platforms. In this view we can see the new signalling centre to the left of the old box, situated on York Road. No. 55014 *The Duke of Wellington's Regiment* displays the head code 1A07, this was the 07.25 Newcastle – Kings Cross. Here it backs onto the stock to form the 1S42 16.00 to Edinburgh.
Photo: G. H. Taylor © Transport Treasury

Above: 8th October 1975 •
An unidentified Cravens DMU has arrived at Kings Cross York Road with the 13.40 from Hertford North.
Photo: G. H. Taylor © Transport Treasury

Left: 3rd February 1978 • No. 31202 at Kings Cross with the 13.30 departure to Cambridge. There's plenty of steam escaping from various coaches on what was a very chilly day.
Photo: G. H. Taylor © Transport Treasury

Left: 4th August 1980 • A stranger at Kings Cross stabled near the Power Box. No. 37072, at this time based in the north east at Thornaby depot, awaits its next duty. Kings Cross men were not trained on Class 37s so it would have been worked in by a visiting crew from one of either Peterborough, Doncaster, York or Newcastle (Gateshead) depots, who signed these locos.
Photo: © Roger Geach

Left: **8th October 1975** • Pictured at Farringdon, a four coach Cravens DMU is heading for Moorgate on an empty coaching stock move. *Photo: G. H. Taylor © Transport Treasury*

Below: **11th February 1977** • Track relaying and reconstruction is ongoing at Kings Cross. No. 31227 of Finsbury Park is working within the possession on a short engineers train. The signal gantry is new and yet to be fitted out with the new colour light signals. *Photo: G. H. Taylor © Transport Treasury*

Left: **9th March 1979** • Midland Region, Toton-based Class 45/1 No. 45129 was another visitor at Kings Cross having worked the 10.20 from Newcastle in the morning rather than a Deltic or Class 47. This was a very cold day with lots of disruption. Class 45 locomotives were very rare at Kings Cross and I was surprised to see this one here. *Photo: © Roger Geach*

8th October 1975 • At Farringdon we see Class 31 No. 31187 of Finsbury Park taking a train of 'block enders' empty stock to Moorgate.
Photo: G. H. Taylor © Transport Treasury

5th September 1978 • On the stops at Kings Cross at around 09.45 hrs. The station roof is being repainted and cleaned while No. 55022 *Royal Scots Grey* has arrived with an unidentified service, alongside is a class 47 hauled service. On the right EMU No. 312024 has arrived from Royston. *Photo: © Roger Geach*

THE HEADSHUNT

In future issues our aim is to bring you many differing articles about the L.N.E.R., its constituent companies and the Eastern and North Eastern regions of British Railways. We hope to have gone some way to achieving this in previous issues.

Eastern Times welcomes constructive comment from readers either by way of additional information on subjects already published or suggestions for new topics that you would like to see addressed. The size and diversity of the L.N.E.R., due to it being comprised of many different companies each with their differing ways of operating, shows the complexity of the subject and we will endeavour to be as accurate as possible but would appreciate any comments to the contrary.

We want to use this final page – The Headshunt – as your platform for comment and discussion so please feel free to send your comments to: tteasterntimes@gmail.com or write to Eastern Times, Transport Treasury Publishing Ltd., 16 Highworth Close, High Wycombe HP13 7PJ.

TALES OF A B17 SPOTTER

I read with interest the article from John Butcher in issue 2.

I am a native of Sudbury, one station down from John on the Stour Valley line. My childhood home in Cross Street backed onto the water meadows providing me with great views of the trains en route or from Melford. In the early 50s my Gran used to take me down to Sudbury station to watch the shunting. Generally a J15 or J17 would be parked at the yard throat next to the footbridge on the path from the station to Cornard Road. These would periodically shuffle across Great Eastern Road on the sidings serving Wheelers timber yard, granaries and coal merchants.

In later life, when using the line to commute, I asked Cyril, the resident platelayer, whether I was mistaken because there was a waggon turntable on each line just inside the yard boundary. He explained that wooden chocks would be hammered in to fix the turntable for the loco to traverse. Being remote from authority in London had its advantages. In later years diesel railcars would often have a waggon in tow. Haslams fish merchants had taken over the original passenger terminus in the goods yard and the railcar would run around the Insulfish van or other waggons and shunt it into the old platform before knocking other wagons around to order. Between times the shunters had a machine much like a rail mounted motor mower to shunt individual wagons. One man pushing and another walking alongside with a shunter pole to pin down the brakes.

Although I can recall riding behind ancient J15s and E4s in ancient coal dusted compartment coaches in the early 50s, by the time I began taking a real interest in the late 50s and watched the trains from the comfort of our back garden, it was mainly B17s, one of which was paired with a black tender possibly borrowed from a K1 or K3 consigned to the cutters. Larger locos such as Riddles and Stanier 2-8-0s still plodded through on freights, but Class 15s and 31s were increasingly in evidence. One 31 was painted in desert sand and this and other class members seemed to very swiftly displace steam on the Summer services from the Midlands to Clacton and Walton seaside resorts. The local Co-operative Societies ran annual specials to Clacton and Walton from the Stour and Colne Valley lines. I can recall them being hauled by sparkling clean B12s hauling a mix of Gresley and Thompson stock. The Colne Valley portion would be attached and detached at Chappel and on the return trip stewards made strenuous efforts to ensure trippers were in the correct part of the train before it divided at Chappel. There were dark rumours amongst railway staff that on one occasion a Britannia was dispatched down the line. How that got through the sharply curved platforms at Sudbury I have no idea.

I only have one memory of the railbuses. My sister and I were sitting on the overbridge at Brundon. This afforded a panoramic view of trains rounding Sudbury and heading straight across the meadows towards Melford. The railbus came bouncing along the jointed track and ground to a halt in front of us, the driver disembarked and walked around the unit examining the mechanical gubbins. He dealt something a sharp blow with a large hammer, climbed aboard and rattled off to Melford.

As an afterthought, I now live in Station Road. In the late 90s the Town Council decided to resurface the lane behind my house. In the process they uncovered the complete siding running along the lane still set in GER oval chairs. This had served a granary and was intended to reach the gas works at the bottom of Edgeworth Road. This never happened but a workshop at the far end of the lanes was built with a curved corner as if to accommodate the left turn to be taken by the track.

And finally, regarding L.N.E.R. concrete signal posts there was a massive example bearing the starting signal on the Down platform at Sudbury. This was designed to be seen from Melford trains over the station roofs and footbridge on the sharp curve through the station. When the station was demolished the post was felled, a digger excavated a trench alongside it and pushed it in. As far as I am aware it is now under the new station car park!

Hope this is of interest

Regards, Mike Davies

SUDBURY – PART 1

I hope this might interest you.

The attached photo shows a Derby lightweight unit leaving Sudbury in October 1968 around one year after closure of the Stour Valley line past Sudbury to Haverhill and Cambridge. The GER through station built on the extension is just out of sight to the left behind the footbridge. Track demolition has just begun with the passing loop to the left and goods yard headshunt to the right having been ripped up. You can see a stack of track panels sitting on the loading dock under the footbridge. However, opposite Sudbury Goods junction signal box, a new line has been laid and ballasted to passenger standards using redundant materials. This curves around into the derelict site of the goods yard heading towards the church tower of St. Peter's in the town centre. This track has been relaid on the original 1848 alignment of the branch into the terminal station, adjacent to the town centre, which, at this point, was still complete with platform and buildings having been relegated to goods traffic when the Stour Valley line opened, and the through station was built. The advertised plan was to relocate branch services into this very convenient location now that the reason for the through station had been removed. However, the plan fell by the wayside, probably through failure to agree funding support with the Council (who not unreasonably wanted a five year guarantee of services remaining on the branch to protect their investment) and the track was quietly removed. A superstore and car park now cover the site and new electro diesel dual mode 3-car units provide an hourly, well patronised service from that loading dock stacked with redundant track panels. The signal box was demolished and nature has reclaimed the trackbed. Another missed opportunity as BR managers seemingly fell over themselves to close lines in the Beeching era.

SUDBURY – PART 2

Dear Peter,

Thanks for the reply. I must congratulate you on the new magazine. A really interesting well presented mix of subject matter. I am seriously tempted to sample the other 3!

Composing the emails I sent you got me thinking about a fairly unique aspect of Sudbury station. Suffolk is not well endowed with railway tunnels. I believe that, aside from Sudbury, the only other is at Ipswich on the main line to Norwich. The tunnel at Sudbury was on a short branch from the goods yard headshunt to the Great Eastern Railway chalk pit. The branch entered a deep cutting and then tunnelled under Cornard Road before emerging in the bottom of the pit, terminating in a fan of sidings by the lime kilns. At the time (and into the 1960s) lime was a profitable traffic providing an important fertiliser for heavy clay fields. I think the siding was removed in the inter war period. I can remember as a child looking over the parapet of the bridge into a deep grassy cutting full of chickens. The pit end of the tunnel had disappeared by then under a new CAV diesel injector factory. One might question why the GER opened this quarry in the town centre. In fact there are several, as the quarry covered a relatively small area and chalk was mined from galleries tunnelled into the rock. Sudbury is honeycombed with these workings and, to digress, my grandfather often told me of driving horses and a cart from the GER pit under the town cemetery to emerge in a pit on the other side of town!

These two photos from a town historical site show lime wagons on the pit siding which runs through the gate lower right towards the tunnel. How neat and tidy everything is. Immediately behind the photographer's left shoulder is Sudbury Goods Junction box. Behind that a short siding led down to a quay on a spur of the Stour Navigation. This had opened from Sudbury to the sea in 1713 and, until the coming of the railway, bricks from Sudbury were transhipped at Mistley port and transported to London by sailing barge. Subsequently they were transhipped at Sudbury from the barges servicing the brick works. All this traffic disappeared by around 1916 when the last barge sailed to the sea and most were then sunk in the canal at Sudbury. Some years ago one was raised and restored and is now moored in the old canal basin.

The second photo is taken from inside the tunnel looking North into the pit. I doubt this lightweight track ever saw a loco! Looks like a job for the station shunting horse.

Regards, Mike Davies

D5512 AT MANOR WAY LEVEL CROSSING

Dear Editor

I enjoyed Dave Brennand's Stratford article in issue 3, especially the centre spread image of Brush Type 2 D5512 at Manor Way level crossing in 1969. It brought to my mind a photograph I took of that same locomotive 10 years earlier, on 17th March 1959, during one of my frequent 'bunks' of Stratford shed when I was an enthusiastic East End trainspotter. What has always intrigued me about my photograph is the huge spanner one of the railwaymen pictured beside the diesel is carrying. It appears to be far too large for everyday use, even in a locomotive context, so, as I am no expert on railway engineering, perhaps an Eastern Times reader could elucidate.

Regards,
Geoff Courtney

SUDBURY MYSTERY

The attached unattributed photo was posted on a Sudbury Town history discussion site.

It comprises a panoramic view South from St Peters Church tower in the town centre. In the immediate foreground is Wheelers timber yard which was Rail served until the 60s. Above its roof in the middle distance can be seen the original Eastern Union terminal station c1849 and engine shed together with parts of the extensive goods yards. When the Great Eastern Railway extended the branch around Sudbury to Cambridge a new substantial station was built and the terminus relegated to goods traffic. In the middle distance is the imposing Olivers Brewery. This was taken over by Greene King and subsequently demolished.

Behind Olivers Brewery to the left the tall shape of Sudbury Goods Junction box can be discerned. The line of trees in front of the box heading left towards Cornard Road marks the deep cutting leading to a tunnel under the road providing rail access into the Railway owned chalk pit.

To the right of the signal box passing behind the brewery roof into the Meadows can be seen the formation of the spur line serving a canal quay on the Stour Navigation. A branch off of the river further upstream dived through a deep cutting into the bottom of the brick pits in the hillside facing Sudbury. After the coming of the railway bricks were transhipped at Sudbury rather than sailing down to Mistley for loading onto seagoing vessels. Many important buildings in London including The Albert Hall and Liverpool Street station were built with Sudbury bricks.

Now comes the puzzle. A sinuous track can be seen heading up the road from the Goods station past the church. This is a deceptively steep incline hence the sinuous course. At first glance I discounted this as a waggon track but on reflection the roads in the town centre were tarmacked so they did not become rutted. The area in front of the photographer was and is an important Junction where routes from Ipswich, Bury St Edmunds, Colchester and London converge yet there are no other waggon tracks. If a Waggoner cut up the roads to this extent he would have found the wrath of the Council falling around his ears. Yet Sudbury never had a tramway. If this is a rail line it does not appear on old O/S maps. The tracks are heading in the general direction of large granaries and chalk pits to the East of the town centre. Can anyone shed light on this?

Regards, Mike Davies